Student Solutions Manual

for use with

Mathematics in Our World

Allan G. Bluman

Professor Emeritus
Community College of Allegheny County

Prepared by
Carrie Green
creativemath

 **Higher Education**

Boston   Burr Ridge, IL   Dubuque, IA   Madison, WI   New York   San Francisco   St. Louis
Bangkok   Bogotá   Caracas   Kuala Lumpur   Lisbon   London   Madrid   Mexico City
Milan   Montreal   New Delhi   Santiago   Seoul   Singapore   Sydney   Taipei   Toronto

The **McGraw·Hill** Companies

Student Solutions Manual for use with
MATHEMATICS IN OUR WORLD
ALLAN G. BLUMAN

Published by McGraw-Hill Higher Education, an imprint of The McGraw-Hill Companies, Inc.,
1221 Avenue of the Americas, New York, NY 10020. Copyright © 2005 by The McGraw-Hill
Companies, Inc. All rights reserved.

This book is printed on acid-free paper.

1 2 3 4 5 6 7 8 9 0 QPD QPD 0 9 8 7 6 5 4

ISBN 0-07-250418-8

www.mhhe.com

Table of Contents

1   PROBLEM SOLVING

Exercise Set 1-1

1. 1   2   4   7   11   16   22   29

   +1 +2 +3 +4 +5 +6 +7 +8

The next number is 37.

3. 10   20   11   18   12   16   13   14   14   12   15

  +10 +(−9) +7 +(−6) +4 +(−3) +1 +0 +(−2) +3 +(−5)

The next number is 10.

5. 100 99   97   94   90   85 79

   −1 −2 −3 −4 −5 −6 −7

The next number is 72.

7. The line through the circle is horizontal, vertical, horizontal, vertical, and then horizontal. The first two circles have no shading, the third and fourth circles have dark shading on top and then on the left, and the fifth circle has

light shading on top. We could reasonably expect the next figure to be .

9. $5 + 13 + 17 = 35$, which is odd.

11. $5^2 \div 2 = 12.5$

13. Approach: Induction

Original number:	10	50
Double the number:	$10(2) = 20$	$50(2) = 100$
Subtract 20:	$20 - 20 = 0$	$100 - 20 = 80$
Divide by 2:	$0 \div 2 = 0$	$80 \div 2 = 40$
Subtract the original number:	$0 - 10 = -10$	$40 - 50 = -10$
Result:	-10	-10

Conjecture: The final answer is −10.

Approach: Deduction

Select a number:	x
Double the number:	$2x$
Subtract 20:	$2x - 20$
Divide by 2:	$(2x - 20) \div 2 = x - 10$
Subtract the original number:	$x - 10 - x = -10$
Result:	-10

The final answer is always −10.

15. Approach: Induction

Original number:	4	13
Add 50:	$4 + 50 = 54$	$13 + 50 = 63$
Multiply by 2:	$54(2) = 108$	$63(2) = 126$
Subtract 60:	$108 - 60 = 48$	$126 - 60 = 66$
Divide by 2:	$48 \div 2 = 24$	$66 \div 2 = 33$
Subtract the original number:	$24 - 4 = 20$	$33 - 13 = 20$
Result:	20	20

Conjecture: The final answer is 20.

Approach: Deduction

Select a number:	x
Add 50:	$x + 50$
Multiply by 2:	$2(x + 50) = 2x + 100$
Subtract 60:	$2x + 100 - 60 = 2x + 40$
Divide by 2:	$(2x + 40) \div 2 = x + 20$
Subtract the original number:	$x + 20 - x = 20$
Result:	20

The final answer is always 20.

17. Answers will vary.

19. It may be impossible or impractical to verify the conclusion for *all* applicable cases.

21. It cannot be done. Four colors are the most that is needed.

23. The numbers that are not 1 in the table are found by adding the two numbers directly above left and above right of each number. The next numbers in the table are

$1 + 5 = 6$
$5 + 10 = 15$
$10 + 10 = 20$
$10 + 5 = 15$
$5 + 1 = 6.$

Exercise Set 1-2

1. Step 1   *Understand the problem.* The coins are worth a total of $4.25. There are quarters and half dollars in the pile and two more quarters than half dollars. A quarter is worth $0.25 and a half dollar is worth $0.50. Find how many quarters and half dollars will give $4.25.

Step 2   *Devise a plan to solve the problem.* Make a list of possible combinations of quarters and half-dollars and see if the sum is $4.25.

Step 3   *Carry out the plan to solve the problem.*

Quarters	Half Dollars	Amount
3	1	$1.25
4	2	$2.00
5	3	$2.75
6	4	$3.50
7	5	$4.25 ← correct

Answer: 7 quarters and 5 half dollars

Step 4   *Check the answer.*
$7 \times \$0.25 + 5 \times \$0.50 = \$4.25.$

3. Step 1   *Understand the problem.* Admission for adults was $1.00 and admission for children was $0.50. 30 tickets were sold for a total of $25.00. Find how many adult tickets and children tickets sold will give $25.00. The answer is the number of adult tickets sold.

Step 2   *Devise a plan to solve the problem.* Make a list of possible combinations of adult and child tickets and see if the total dollar amount is $25.00.

Step 3   *Carry out the plan to solve the problem.*

Adults	Children	Dollar Amount
10	20	$20.00
12	18	$21.00
16	14	$23.00
18	12	$24.00
19	11	$24.50
20	10	$25.00

Answer: 20 adult tickets were sold.

Step 4   *Check the answer.*
$20 \times \$1.00 + 10 \times \$0.50 = \$25.00$

5. **Step 1**   *Understand the problem.* There are chickens and cows in a barnyard. A chicken has one head and 2 legs. A cow has one head and four legs. There are 12 heads and 38 legs. Find how many chickens and cows there are.

Step 2   *Devise a plan to solve the problem.* Make a list of possible combinations of numbers of chickens and cows and see if there are 38 legs.

Step 3   *Carry out the plan to solve the problem.*

Chickens	Cows	Number of Legs
3	9	42
4	8	40
5	7	38 ← correct

Answer: 5 chickens and 7 cows

Step 4   *Check the answer.*
$5 + 7 = 12$ and $5(2) + 7(4) = 38$

7. **Step 1**   *Understand the problem.* In a two-digit number, the sum of the two digits is 7. The number minus 9 is the original number with the digits reversed. Find the number.

Step 2   *Devise a plan to solve the problem.* Make a list of two-digit numbers where the sum of the digits is 7. See which number minus 9 is a number with the digits reversed.

Step 3   *Carry out the plan to solve the problem.*

Original Number	Number minus 9
16	7
25	16
34	25
43	34 ← correct
52	43
61	52
70	61

Answer: The number is 43.

Step 4   *Check the answer.* $34 + 9 = 43$

9. **Step 1**   *Understand the problem.* Find two numbers for which one number plus 7 equals the other number and one number plus the other number equals 23.

Step 2   *Devise a plan to solve the problem.* Make a list of numbers where one number is 7 more than another number. Find the pair whose sum equals 23.

Step 3   *Carry out the plan to solve the problem.*

First Number	Second Number	Sum
4	11	15
6	13	20
7	14	21
8	15	23

Answer: The numbers are 8 and 15.

Step 4   *Check the answer.* $8 + 7 = 15$; $8 + 15 = 23$

11. **Step 1**   *Understand the problem.* When we add 28 to Mark's current age, the result is five times his current age. Find Mark's current age.

Step 2   *Devise a plan to solve the problem.* Make a list of pairs of ages where one age is 28 more than the other age. Find the pair where one age is five times the other age.

Step 3   *Carry out the plan to solve the problem.*

Current Age	Age + 28
3	31
5	33
6	34
7	35 ← correct

Answer: Mark is 7 years old.

Step 4   *Check the answer.* 7 + 28 = 35;
7 × 5 = 35

13. Step 1   *Understand the problem.* Pete's current age equals Lashanna's current age times 2. Pete's age plus 5 added to Lashanna's age plus 5 equals 37. Find Pete's and Lashanna's ages.

Step 2   *Devise a plan to solve the problem.* Make a list of pairs of ages where Pete's age is twice Lashanna's age. Add 5 to each age and see which pair adds up to 37.

Step 3   *Carry out the plan to solve the problem.*

Lashanna's Age	Pete's Age	Lashanna's Age + 5	Pete's Age + 5
3	6	8	11
5	10	10	15
7	14	12	19
8	16	13	21
9	18	14	23 ↑ correct

Answer: Lashanna is 9 years old and Pete is 18 years old.

Step 4   *Check the answer.* 9 × 2 = 18;
9 + 5 + 18 + 5 = 37

15. Step 1   *Understand the problem.* It takes 2 hours to cut the grass, where Sam works for $1\frac{1}{2}$ hours and Pete works $\frac{1}{2}$ hour. The total amount paid is $60. Find the amounts Sam and Pete received so that the total amount is $60.

Step 2   *Devise a plan to solve the problem.* We must find the hourly wage paid for cutting the grass. It takes 2 hours to cut, so divide the total amount paid by 2 to find the hourly amount.

Step 3   *Carry out the plan to solve the problem.* Divide $60 by 2: $60 ÷ 2 = $30 per hour. If Sam works $1\frac{1}{2}$ hours then he receives $1\frac{1}{2} \times \$30 = \$45$. If Pete works $\frac{1}{2}$ hour then he receives $\frac{1}{2} \times \$30 = \$15$.

Step 4   *Check the answer.* Sam receives $45 and Pete receives $15, so the total amount paid was $45 + $15 = $60.

17. Step 1   *Understand the problem.* Given the weights of one bag of apples and one bag of pears, find the weights of three bags of apples and six bags of pears. Then find the total weight for the nine bags of fruit.

Step 2   *Devise a plan to solve the problem.* Multiply the weight of one bag of apples by 3 and multiply the weight of one bag of pears by 6. Then add these products to find the total weight.

Step 3   *Carry out the plan to solve the problem.* Three bags of apples weight 3 × 8 = 24 pounds. Six bags of pears weight 6 × 10 = 60 pounds. The total weight of the bags of fruit is 24 + 60 = 84 pounds.

Step 4   *Check the answer.*
3(8) + 6(10) = 24 + 60 = 84.

19. Step 1   *Understand the problem.* A person has $1624 per month to pay bills and cover living expenses. Some of those expenses are $256 for food, $125 for gasoline, and $150 for utilities. Find how much money is left for other expenses.

Step 2   *Devise a plan to solve the problem.* Add the known expenses and subtract the total from the amount of money available for the month. The result is the amount of money left to pay other bills.

Step 3 *Carry out the plan to solve the problem.* Add the costs for food, gasoline, and utilities: $256 + $125 + $150 = $531. Subtract the total known expenses from the amount of money available: $1624 − $531 = $1093. $1093 is the amount left for other monthly expenses.

Step 4 *Check the answer.* $1093 + $150 + $125 + $256 = $1624

21. **Step 1** *Understand the problem.* There are 26 bags of pretzels. Each bag contains 1650 calories. Find the total calories in all the bags of pretzels.

Step 2 *Devise a plan to solve the problem.* Multiply the number of bags by the number of calories per bag. This gives the total calories in all the bags.

Step 3 *Carry out the plan to solve the problem.* Multiply 26 bags by 1650 calories per bag: 26 × 1650 = 42,900 calories. There are 42,900 calories in all the bags of pretzels.

Step 4 *Check the answer.* 42,900 ÷ 26 = 1650

23. **Step 1** *Understand the problem.* The loan is for a total of $12,381. The loan is to be paid once a month for 5 years. Find the amount of each monthly payment.

Step 2 *Devise a plan to solve the problem.* First we must find the total number of payments to be made over 5 years, using the fact that there are 12 months per year. Then we can divide the loan amount by the number of payments that will be made to find the monthly payment amount.

Step 3 *Carry out the plan to solve the problem.* Find the total number of payments by multiplying 5 years by 12 months per year: 5 × 12 = 60, so 60 payments will be made. Then, divide the loan amount by the number of payments: $12,381 ÷ 60 = $206.35. The monthly payment should be $206.35.

Step 4 *Check the answer.* $206.35 × 12 months = $2476.20 paid per year; $2476.20 × 5 years = $12,381.

25. **Step 1** *Understand the problem.* Find the total cost to buy five maple saplings, eight birch saplings, and three oak saplings if maple saplings cost $32 each, birch saplings cost $20 each, and oak saplings cost $25 each.

Step 2 *Devise a plan to solve the problem.* Multiplyh the cost per sapling by the number of each type of sapling to be purchased. Then add the costs for each together to get the total cost.

Step 3 *Carry out the plan to solve the problem.* Find the cost for each type of sapling: maple: 5 × $32 = $160 birch: 8 × $20 = $160 oak: 3 × $25 = $75 Add these costs together: $160 + $160 + $75 = $395 The total cost for the saplings is $395.

Step 4 *Check the answer.* (5 × 32) + (8 × 20) + (3 × 25) = 395

27. **Step 1** *Understand the problem.* A person drives 864 miles for business. The person will be reimbursed 26.5¢ per mile. Find the amount the person will be reimbursed.

Step 2 *Devise a plan to solve the problem.* Multiply the number of miles by the price per mile. This gives the total amount the person will be reimbursed.

Step 3 *Carry out the plan to solve the problem.* Multiply 864 miles by 26.5¢ per mile: 864 × 26.5 = 22,896 cents, or $228.96. The person will be reimbursed $228.96.

Step 4 *Check the answer.* $228.96 ÷ 864 miles = $0.265, or 26.5¢ per mile.

29. **Step 1** *Understand the problem.* The clerk's regular hourly rate is $9.50 per hour for working more than 0 but less than or equal to 40 hours. If the clerk works more than 40 hours the clerk's hourly rate for any hours more than 40 hours is one and one-half times the regular hourly rate. Find how much the clerk earned for working 46 hours in one week.

Step 2 *Devise a plan to solve the problem.* First find the clerk's hourly rate for working more than 40 hours by multiplying the regular hourly rate of $9.50 by $1\frac{1}{2}$. Then, since the clerk worked 40 hours at the regular rate and 6 hours at the other rate, find how much the clerk earned by multiplying each rate by the number of hours worked at that rate. Add the results to get the total amount earned.

Step 3 *Carry out the plan to solve the problem.* Find the clerk's hourly rate for working more than 40 hours:

$9.50 \times 1\frac{1}{2} = $14.25.$ The clerk earns $14.25 per hour for any hours worked over 40 hours. Then the clerk earned $9.50 \times 40 = 380 for the first 40 hours worked and $14.25 \times 6 = 85.50 for the last 6 hours worked. The total amount earned was $380 + $85.50 = $465.50.

Step 4 *Check the answer.*

$$40 \times $9.50 + 6 \times 1\frac{1}{2} \times $9.50 = $465.50$$

31. 1. Understand the problem. Read the problem several times, noting important information and what you are asked to find.

2. Devise a plan to solve the problem. Find an organized and logical method to take the given information and use it to obtain the answer.

3. Carry out the plan to solve the problem. Use the plan to find the answer. Use a different plan if the original plan doesn't work.

4. Check the answer. Use another method, estimation, or reason to check that the answer you obtained is correct.

33. Step 1 *Understand the problem.* The king pays the knight for 6 days of work using a 6-inch gold bar. The king pays the knight an equal amount of the bar each day. The king only makes 2 cuts in the bar. Find how the king was able to pay the knight daily.

Step 2 *Devise a plan to solve the problem.* The gold bar is 6 inches long, and the king pays the knight over 6 days so the king pays the knight in 1-inch pieces each day. Find a way the king can cut the bar using only two cuts so that the king can pay the knight with a 1-inch piece each day. Visualize the payments with a sketch of the gold bar.

Step 3 *Carry out the plan to solve the problem.* The king must pay the knight with 1 inch of the bar at the end of the first day, so the first cut must be made at the 1-inch mark on the bar. At the end of the second day the king must pay the knight with another 1-inch piece, but if he cuts the bar at the original 2-inch mark on the bar the remaining piece will be 4 inches and too big to pay the knight daily without making another cut. The king should cut the remaining 5 inches of the bar at the 3-inch mark so that there is a 3-inch piece and a 2-inch piece. The king will take the 1-inch piece back from the knight and give him the 2-inch piece at the end of the second day. Continuing in this manner, the king can pay the knight an additional 1-inch piece of gold bar for each of the 6 days.

Step 4 *Check the answer.* The king made only 2 cuts, and the knight was paid in 1-inch pieces each day for six days.

35. Step 1 *Understand the problem.* An automobile is worth $18,000 after 1 year. This amount is the value of the automobile after 20% depreciation of its original value. Find the original value of the automobile.

Step 2 *Devise a plan to solve the problem.* The value after 20% depreciation is 80% of the original value. So if $18,000 is 80% of the original value x, solve for x by dividing $18,000 by 80% (or 0.80).

Step 3 *Carry out the plan to solve the problem.* Solve for x:

$18,000 = 80\%(x)$
$18,000 = 0.80x$
$\dfrac{$18,000}{0.80} = x$
$22,500 = x$

The original value of the automobile was $22,500.

Step 4 *Check the answer.*
$22,500(20\%) = $22,500(0.20)$
$= $4,500;$
$22,500 - $4,500 = $18,000.$

37. Step 1 *Understand the problem.* The cost of a computer is covered with $\frac{1}{3}$ of the money from the freshmen, $\frac{1}{2}$ of the money from the sophomores, and $400 from the Student Government Association. Find the cost of the computer.

Step 2 *Devise a plan to solve the problem.* Let x represent the cost of the computer. Then the freshmen paid $\frac{1}{3}x$ and the sophomores paid $\frac{1}{2}x$. Add the amounts contributed by each group and solve for x.

Step 3 *Carry out the plan to solve the problem.*

$$\frac{1}{3}x+\frac{1}{2}x+\$400 = x$$
$$\frac{2}{6}x+\frac{3}{6}x+\$400 = x$$
$$\frac{5}{6}x+\$400 = x$$
$$\$400 = \frac{1}{6}x$$
$$6(\$400) = x$$
$$\$2400 = x$$

The computer cost $2400.

Step 4 *Check the answer.*

$$\frac{1}{3}(\$2400)+\frac{1}{2}(\$2400)+\$400$$
$$= \$800+\$1200+\$400$$
$$= \$2400$$

Exercise Set 1-3

1. In the number 2861, the 8 is the digit being rounded. Since the digit to the right is 6, 1 is added to the 8 and the digits 6 and 1 are replaced by zeros. The rounded number is 2900.

3. In the number 3,261,437, the 6 is the digit being rounded. Since the digit to the right is 1, the digit 6 remains the same and the digits to the right are replaced by zeros. The rounded number is 3,260,000.

5. In the number 62.67, the 2 is the digit being rounded. Since the digit to the right is 6, 1 is added to the 2 and the digits to the right are replaced by zeros. The rounded number is 63.00 or 63.

7. In the number 218,763, the 2 is the digit being rounded. Since the digit to the right is 1, the digit 2 remains the same and the digits to the right are replaced by zeros. The rounded number is 200,000.

9. In the number 3.671, the 7 is the digit being rounded. Since the digit to the right is 1, the digit 7 remains the same and the digit 1 is replaced by zero. The rounded number is 3.670 or 3.67.

11. In the number 327.146, the 1 is the digit being rounded. Since the digit to the right is 4, the digit 1 remains the same and the digits 4 and 6 are replaced by zeros. The rounded number is 327.100 or 327.1.

13. In the number 5,462,371, the 6 is the digit being rounded. Since the digit to the right is 2, the digit 6 remains the same and the digits to the right are replaced by zeros. The rounded number is 5,460,000.

15. In the number 272,341, the 2 is the digit being rounded. Since the digit to the right is 7, 1 is added to the 2 and the digits to the right are replaced by zeros. The rounded number is 300,000.

17. In the number 264.97348, the 4 is the digit being rounded. Since the digit to the right is 8, 1 is added to the 4 and the digit 8 is replaced by zero. The rounded number is 264.97350 or 264.9735.

19. In the number 563.271, the 7 is the digit being rounded. Since the digit to the right is 1, the digit 7 remains the same and the digit 1 is replaced by zero. The rounded number is 563.270 or 563.27.

21. Round the cost of each tire to $60. Then $8 \times \$60 = \$480$. The estimated cost of the tires is $480.

23. Round the distance to 240 miles and the speed to 40 miles per hour. Then $240 \div 40 = 6$, so the estimated time is 6 hours.

25. Round the original cost to $180. The discount is 60% of $180, or $108, so the sale price is estimated to be $180 − $108 = $72.

27. Round each price and add:

$$\begin{array}{ll}\$1.89 & \to \quad \$2.00 \\ 1.29 & \to \quad 1.00 \\ 0.89 & \to \quad + 1.00 \\ \hline & \quad \$4.00\end{array}$$

The total estimated cost is $4.00.

29. 86 is close to 81 and 8 is close to 9, so it takes about $81 \div 9 = 9$ hours. Or, since 86 is close to 90 and 8 is close to 9, it takes about $90 \div 9 = 10$ hours. Or, since 86 is close to 88, it takes about $88 \div 8 = 11$ hours.

31. Round $8.75 to $9.00. The person works 40 hours per week times 50 weeks per year or $40 \times 50 = 2000$ hours per year. Then the person earns about $\$9.00 \times 2000 = \$18,000$ per year.

33. Round 24 feet to 25 feet, round 18 feet to 20 feet, and round $5.95 to $6.00. The length of fencing needed is
25 feet + 20 feet + 25 feet + 20 feet = 90 feet.
Then the cost for the fencing is about
$\$6 \times 90 = \$540$.

35. The total area of the yard is $16 \times 16 = 256$ square feet. The number of boxes of grass seed needed is $256 \div 24 = 10\frac{2}{3}$. Then, since each box of grass seed costs $5.95 or about $6, the cost for planting grass is about $10\frac{2}{3} \times \$6 = \$64$.

37. Locate the bar representing the area of Lake Huron and read across to the vertical axis. Since the top of the bar extends to about halfway between 20,000 and 25,000 square miles, we estimate the area to be 23,000 square miles (to the nearest thousand miles).

39. Lake Superior is the largest and Lake Erie is the smallest. We estimate the area of Lake Superior to be about 32,000 square miles and the area of Lake Erie is about 10,000 square miles. So the difference is
$32,000 - 10,000 = 22,000$ square miles.

41. The sector of the graph corresponding to people that feel late morning and early afternoon are the most productive times of day has a percentage of 22%. 22% of 1385 office workers is 0.22 (1385) = 304.7 or about 305 people.

43. Locate the year 1950 halfway between 1940 and 1960 on the horizontal axis and move up to the line on the graph. At this point, move horizontally to the vertical axis. We estimate that 350 billion cigarettes were smoked in 1950.

45. Since $\frac{1}{4}$ inch = 40 miles, find the number of $\frac{1}{4}$-inch parts there are in $3\frac{1}{2}$ inches:
$3\frac{1}{2} \div \frac{1}{4} = \frac{7}{2} \div \frac{1}{4} = \frac{7}{2} \cdot \frac{4}{1} = 14$. Then $14 \times 40 = 560$, so $3\frac{1}{2}$ inches represents 560 miles, and the cities are 560 miles apart.

47. (a) Buying groceries

(b) "Having an idea" of the time needed for a trip to the post office

(c) Figuring the number of gallons of milk produced per cow, per month

49. If the estimate falls within a "reasonable distance" of the exact answer, then the latter has a better (greater) probability of being correct.

51. Answers will vary.

Review Exercises

1. 3   4   6   7   9   10   12   13   15   16

+1 +2 +1 +2 +1 +2 +1 +2 +1 +2 +1 +2

The next three numbers are 18, 19, and 21.

3. 4   z   16   w   64   t   256

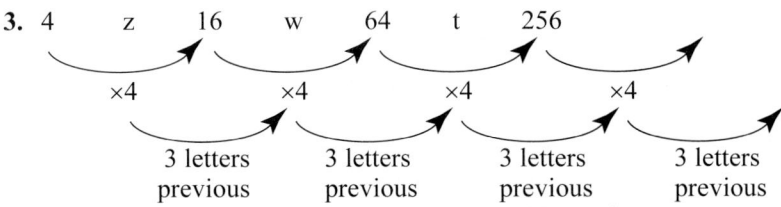

×4   ×4   ×4   ×4

3 letters   3 letters   3 letters   3 letters
previous   previous   previous   previous

The next three items in the sequence are q, 1024, and n.

5. There are four circles with solid-circle centers then two squares with open-circle centers. On the outsides of the figures there are little lines at the top, right, bottom, left, top, and then right. We could reasonably expect the next

figure to be [○] .

7. $5(7)(11) = 385$, which is odd.

9. Approach: Induction

Original number:	2	12
Add 6:	$2 + 6 = 8$	$12 + 6 = 18$
Divide by 2:	$8 \div 2 = 4$	$18 \div 2 = 9$
Add 10:	$4 + 10 = 14$	$9 + 10 = 19$
Result:	$14 = \frac{1}{2}(2) + 13$	$19 = \frac{1}{2}(12) + 13$

Conjecture: The final answer is 13 more than $\frac{1}{2}$ of the original even number.

Approach: Deduction

Select a number:	x
Add 6:	$x + 6$
Divide by 2:	$\frac{1}{2}(x + 6) = \frac{1}{2}x + 3$
Add 10:	$\frac{1}{2}x + 3 + 10 = \frac{1}{2}x + 13$
Result:	$\frac{1}{2}x + 13$

The final answer is always 13 more than $\frac{1}{2}$ of the original even number.

11. Cindy gave away all the fish except nine of them. That means, of the 32 fish she started with, she had 9 fish left.

13. Because person's weight while standing on one foot is not supported by anything else, the person's weight on the scale does not change when the person steps on the scale with both feet. So the person still weighs 110 pounds.

15. **Step 1** *Understand the problem.* Currently, Harry is 10 years old and Bill's age is two times Harry's age. Find Bill's age when Harry is 20.

Step 2 *Devise a plan to solve the problem.* Find Bill's current age by multiplying Harry's current age by 2. Then since Harry will be 20 in 20 − 10 years, add 10 to Bill's current age to find Bill's age when Harry is 20.

Step 3 *Carry out the plan to solve the problem.* Find Bill's current age: 2 × 10 = 20 years old. Add 10 to Bill's current age: 20 + 10 = 30 years old. When Harry is 20, Bill will be 30.

Step 4   *Check the answer*. Bill's current age, 20, is twice Harry's current age. 10 years from now, Harry will be $10 + 10 = 20$ and Bill will be $20 + 10 = 30$.

17.   Step 1   *Understand the problem*. A baseball glove costs two times the amount a baseball costs. The total cost is $20.00. Find the cost of the baseball.

Step 2   *Devise a plan to solve the problem*. Make a list of possible combinations of prices for a baseball and a glove and see if the total cost is $20.00.

Step 3   *Carry out the plan to solve the problem*.

Baseball	Glove	Total	
$5.00	$10.00	$15.00	
$6.00	$12.00	$18.00	
$6.50	$13.00	$19.50	
$6.65	$13.30	$19.95	
$6.67	$13.34	$20.01	← correct

The baseball costs approximately $6.67.

Step 4   *Check the answer*. The glove costs approximately $2(\$6.67) = \$13.34$. The total cost for both is then $\$6.67 + \$13.34 = \$20.01$, which is about $20.00.

19.   Step 1   *Understand the problem*. We want to find Harriet's age now, given that 10 years from now she will be three times as old as she was 10 years ago.

Step 2   *Devise a plan to solve the problem*. Make a list of possible combinations of ages for Harriet and see if a pair satisfies the conditions in the problem.

Step 3   *Carry out the plan to solve the problem*.

Harriet's Age Now	Harriet's Age 10 Years Ago	Harriet's Age 10 Years From Now	Quotient	
15	5	25	5	
18	8	28	$\frac{28}{8} = \frac{7}{2}$	
20	10	30	3	← correct

Harriet is 20 years old now.

Step 4   *Check the answer*. 10 years ago Harriet was $20 - 10 = 10$ years old. 10 years from now Harriet will be $20 + 10 = 30$ years old, which is three times 10, Harriet's age 10 years ago.

21.   Step 1   *Understand the problem*. Babs's age plus Debbie's age plus Jack's age equals 20. Babs's age equals Debbie's age plus 9. Jack's age equals Babs's age minus 7. Find Debbie's age.

Step 2   *Devise a plan to solve the problem*. Make a list of combinations of ages for Babs, Debbie, and Jack and see if a combination satisfies the conditions in the problem.

Step 3   *Carry out the plan to solve the problem*.

Babs's Age	Debbie's Age	Jack's Age	Total	
14	5	7	26	
13	4	6	23	
12	3	5	20	← correct

Debbie is 3 years old.

Step 4 *Check the answer.* Debbie is 3 years old, Babs is $3 + 9 = 12$ years old, and Jack is $12 - 7 = 5$ years old. The sum of their ages is $3 + 12 + 5 = 20$.

23. **Step 1** *Understand the problem.* Harry and Bill each have some baseball cards. If Bill gives Harry one card, they will each have the same number of cards. If Harry gives Bill one card, then Bill will have two times as many cards as Harry. Find the number of cards each person has.

Step 2 *Devise a plan to solve the problem.* Make a list of possible numbers of cards Bill and Harry have. Find the pair that satisfy both conditions.

Step 3 *Carry out the plan to solve the problem.*

Bill	Harry	
4	2	
5	3	
6	4	
7	5	← correct

Answer: Bill has 7 cards and Harry has 3 cards.

Step 4 *Check the answer.* If Bill gives Harry one card, then they each have 6 cards. If Harry gives Bill one card, then Bill has 8 cards and Harry has 4 cards so Bill has double the number Harry has.

25. There are three sizes of triangles in the figure. There are 9 of the smallest triangles, 3 triangles each made up of 4 of the smallest triangles, and 1 triangle made up of all of the smallest triangles. Thus, there are $9 + 3 + 1 = 13$ triangles in the figure.

27. **Step 1** *Understand the problem.* Find two numbers that when added equal 20 and when subtracted equal 5.

Step 2 *Devise a plan to solve the problem.* Make a list of pairs of numbers that add to 20. Find the pair whose difference is 5.

Step 3 *Carry out the plan to solve the problem.*

First Number	Second Number	
5	15	
6	14	
7	13	
8	12	
7.5	12.5	← correct

The numbers are 7.5 and 12.5.

Step 4 *Check the answer.* $7.5 + 12.5 = 20$; $12.5 - 7.5 = 5$

29. **Step 1** *Understand the problem.* Mr. Taylor invested part of $1000 in something that earned 8% simple interest and the rest of the $1000 in something that earned 6% simple interest. After 1 year, he had earned $76.00 in interest. Find the amount invested at each rate.

Step 2 *Devise a plan to solve the problem.* Make a list of possible combinations of investments that add up to $1000. Find the combination that gives $76.00 in simple interest. Use the fact that simple interest equals the principal times the rate times the length of time.

Step 3 *Carry out the plan to solve the problem.*

Amount Invested at 8%	Amount Invested at 6%	Interest from 8% Investment	Interest from 6% Investment
$200	$800	$16	$48
$400	$600	$32	$36
$600	$400	$48	$24
$700	$300	$56	$18
$800	$200	$64	$12 ↑ correct

Mr. Taylor invested $800 at 8% and $200 at 6%.

Step 4 *Check the answer.*
$800 + $200 = $1000;
($800)(0.08)(1) + ($200)(0.06)(1)
$= $64 + $12 = $76.

31. In the number 186.75, the 6 is the digit being rounded. Since the digit to the right is 7, 1 is added to the 6 and the digits to the right are replaced by zeros. The rounded number is 187.00 or 187.

33. In the number 0.6314, the 6 is the digit being rounded. Since the digit to the right is 3, the digit 6 remains the same and the digits to the right are replaced by zeros. The rounded number is 0.6000 or 0.6.

35. Round the cost of each lawnmower to $330.00. Then 4 × $330.00 = $1320.00. The estimated cost of the lawnmowers is $1320.

37. The sector of the graph corresponding to pollutants released in the air has a percentage of 60%. 60% of 1953 million pounds is 0.60(1953 million) = 1171.8 million or about 1172 million pounds.

39. Locate the year 1988 between 1985 and 1990 on the horizontal axis and move up to the line on the graph. At this point, move horizontally to the vertical axis. We estimate that the average weekly salary in 1988 was about $350.

41. Since 2 inches = 15 miles, 8 inches represents 4(15 miles) = 60 miles.

Chapter Test

* When using estimation, other correct answers are possible.

1. 2   4   3   6   5   9   8
+2 −1 +3 −1 +4 −1 +5 −1 +6
The next three numbers are 13, 12, and 18.

3. Add enough dots to the last figure given to form a triangle with 4 dots per side. The number below the figure is the number of dots in the figure.

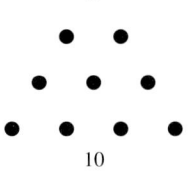

10

5. Approach: Induction

	Original number:	2	7
Add 10:		$2+10=12$	$7+10=17$
Multiply by 5:		$12\times 5=60$	$17\times 5=85$
Add 15:		$60+15=75$	$85+15=100$
Divide by 5:		$75\div 5=15$	$100\div 5=20$
Result:		$15=2+13$	$20=7+13$

Conjecture: The final answer equals the original number plus 13.

Approach: Deduction

Select a number:	x
Add 10:	$x+10$
Multiply by 5:	$5(x+10)=5x+50$
Add 15:	$5x+50+15=5x+65$
Divide by 5:	$(5x+65)\div 5=x+13$
Result:	$x+13$

The final answer is equal to the original number plus 13.

7. **Step 1** *Understand the problem.* The worker earns twice as much as the previous day for a total of 10 days of work. The last day the person earned $100.00. Find which day the worker received $25.00.

Step 2 *Devise a plan to solve the problem.* Since the amount earned on a given day is 2 times the amount earned the previous day, the amount earned the previous day is half the amount earned the given day. Starting with $100.00 for the tenth day, work back in time dividing the day's earnings by 2, until you reach the day the worker earned $25.00 for the day.

Step 3 *Carry out the plan to solve the problem.*

Day of Work	Amount Earned	
Tenth	$100.00	
Ninth	$50.00	
Eight	$25.00	← correct

Answer: On the eight day the worker earned $25.00.

Step 4 *Check the answer.* If the worker earned $25.00 on the eighth day, then the worker earned $50.00 on the ninth day and $100.00 on the tenth day.

9. Move one line as shown:

|\/ − || = || or   4 − 2 = 2

This is an equation in Roman numerals: 4 − 2 = 2.

11. Let x represent the age of the man when he died. Then

Boyhood lasted $\frac{1}{6}x$ years

Beard grew at age $\frac{1}{12}x + \frac{1}{6}x$

Married at age $\frac{1}{7}x + \frac{1}{12}x + \frac{1}{6}x$

Son born at man's age $5 + \frac{1}{7}x + \frac{1}{12}x + \frac{1}{6}x$

Son died at man's age $\frac{1}{2}x + 5 + \frac{1}{7}x + \frac{1}{12}x + \frac{1}{6}x$

Man died at age

$$x = 4 + \frac{1}{2}x + 5 + \frac{1}{7}x + \frac{1}{12}x + \frac{1}{6}x$$

$$x = \frac{25}{28}x + 9$$

$$\frac{3}{28}x = 9$$

$$x = 84$$

The man was 84 when he died.

13. Let x represent the number. Then $\frac{1}{2}x + \frac{1}{3}x = 10$.

Solve for x: $\frac{1}{2}x + \frac{1}{3}x = 10$

$$\frac{5}{6}x = 10$$

$$x = 12$$

15. Step 1 *Understand the problem.* Two people work for 3 hours and 2 hours, respectively. The total amount earned is $60.00. Find how much each person earned for the work.

Step 2 *Devise a plan to solve the problem.* The total time spent was $2 + 3 = 5$ hours, and the total amount earned was $60.00. Find the hourly rate by dividing then find the amount earned by each person by multiplying by the number of hours worked by each person.

Step 3 *Carry out the plan to solve the problem.*
Hourly rate = $60.00 ÷ 5$ hours
    = $12 per hour

First person earns $3 \times \$12 = \$36$.
Second person earns $2 \times \$12 = \$24$.

Step 4 *Check the answer.* $36 + $24 = $60

17. Step 1 *Understand the problem.* Sam has taken an exam and scored 87%. He wants an average exam score of 90%. Find the score Sam needs on his second exam.

Step 2 *Devise a plan to solve the problem.* Make a list of possible second exam scores. Find the average exam score by adding the first and second exam scores and dividing by two. Find the second exam score that gives an average of 90%.

Step 3 *Carry out the plan to solve the problem.*

Second Exam	Average	
90%	88.5%	
91%	89%	
92%	89.5%	
93%	90%	← correct

Answer: Sam needs a 93% on the second exam.

Step 4 *Check the answer.* $87 + 93 = 180$; $180 ÷ 2 = 90$

19. Think of heights and depths above and below sea level as positive distances from a horizontal line. Add the positive distances to find the total distance. 29,000 feet + 36,400 feet = 65,400 feet

21. Let x be Mark's age. Then Mark's mother is $x + 32$ years old and $x + x + 32 = 66$. Solve for x and find $x + 32$:
$$x + x + 32 = 66$$
$$2x = 34$$
$$x = 17$$
$$x + 32 = 49$$
Mark is 17 years old and his mother is 49.

23. In the number of 1.3752, the 7 is the digit being rounded. Since the digit to the right is 5, 1 is added to the 7 and the digits to the right are replaced by zeros. The rounded number is 1.3800 or 1.38.

25. Locate the year 1980 on the horizontal axis and move up to the line on the graph. At this point, move horizontally to the vertical axis. We estimate that the average number of hours per week was about 35.5 hours.

27. Locate the bar representing Baltimore and read across to the vertical axis. Since the top of the bar extends a little bit above 300, we estimate the number of homicides for Baltimore to be about 310.

14

2 Sets

Exercise Set 2-1

1. {s, t, r, e}

3. {51, 52, 53, 54, 55, 56, 57, 58, 59}

5. {1, 3, 5, 7, 9, 11, 13}

7. {11, 12, 13, 14, ...}

9. {2001, 2002, 2003, ..., 2999}

11. {Monday, Tuesday, Wednesday, Thursday, Friday, Saturday, Sunday}

13. {diamond, club spade, heart}

15. {even natural numbers}

17. {the first four multiples of 9}

19. {letters in Mary}

21. {natural numbers from 100 to 199}

23. $\{x \mid x$ is a multiple of 10$\}$

25. $\{x \mid x \in N$ and $x > 20\}$

27. $\{x \mid x$ is an odd natural number less than 10$\}$

29. There are none.

31. {7, 14, 21, 28, 35, 42, 49, 56, 63}

33. {102, 104, 106, ...}

35. Well defined

37. Well defined

39. Not well defined

41. Not well defined

43. True

45. True

47. True

49. Infinite

51. Finite

53. Infinite

55. Finite

57. Equal and equivalent

59. Neither

61. Equivalent

63. Neither

65. {10 20 30 40}
    {10 20 30 40}
    $\updownarrow \updownarrow \updownarrow \updownarrow$
    {40 10 20 30}

67. {1 2 3 ... 25 26}
    $\updownarrow \updownarrow \updownarrow \quad \updownarrow \updownarrow$
    {a b c ... y z}

69. {Carnival, Royal Caribbean, Princess, Holland America, Norwegian}

71. {Seabourn, Windstar}

73. {'94, '95}

75. $\varnothing$

77. A *set* is a well-defined collection of objects.

79. Equal sets have exactly the *same elements*. Equivalent sets have exactly the *same number* of elements.

81. Each element of one set can be associated (paired) with exactly one element of the other set, and no element in either set is left alone.

83. Yes; $A \cong B$ means A and B have the same number of elements. $A \cong C$ means A and C have the same number of elements. Then B and C have the same number of elements, so $B \cong C$.

85. Answers will vary.

87. Yes

Exercise Set 2-2

1. $\varnothing$; {r}; {s}; {t}; {r, s}; {r, t}; {s, t}; {r, s, t}

3. $\varnothing$; {1}; {3}; {1, 3}

5. { } or $\varnothing$

7. $\varnothing$; {5}; {12}; {13}; {14}; {5, 12}; {5, 13};
   {5, 14}; {12, 13}; {12, 14}; {13, 14},
   {5, 12, 13}; {5, 12, 14}; {5, 13, 14};
   {12, 13, 14}; {5, 12, 13, 14}

9. $\varnothing$; {1}; {10}; {20}; {1, 10}; {1, 20}; {20, 20}

11. $\varnothing$

13. None

15. True

17. False; it is not a subset.

19. False; { } has no elements.

21. False; $\subset$ or $\subseteq$ should be used with subsets.

23. True

25. $2^3 = 8$

27. $2^0 = 1$

29. $2^2 = 4$

31. {10, 30, 40, 50, 60, 70, 90}

33. {20, 40, 60, 80, 100}

35. $B \cup C = \{20, 30, 40, 50, 60, 80, 100\}$
    $\overline{A} = \{20, 40, 60, 80, 100\}$
    $\overline{A} \cap (B \cup C) = \{20, 40, 60, 80, 100\}$

37. $A \cup B = 10, 20, 30, 40, 50, 60, 70, 80, 90, 100\}$
    $\overline{A \cup B} = \varnothing$
    $\overline{(A \cup B)} \cap C = \varnothing$

39. $B \cup C = \{20, 30, 40, 50, 60, 80, 100\}$
    $\overline{A} = \{20, 40, 60, 80, 100\}$
    $(B \cup C) \cap \overline{A} = \{20, 40, 60, 80, 100\}$

41. {b, d}

43. {a, c, e, h}

45. $\overline{R} = \{a, b, c, d, h\}$
    $\overline{P} = \{a, c, e, h\}$
    $\overline{R} \cap \overline{P} = \{a, c, h\}$

47. $Q \cup P = \{a, b, c, d, f, g\}$
    $\overline{Q \cup P} = \{e, h\}$
    $\left(\overline{Q \cup P}\right) \cap R = \{e\}$

49. $P \cup Q = \{a, b, c, d, f, g\}$
    $P \cup R = \{b, d, e, f, g\}$
    $(P \cup Q) \cap (P \cup R) = \{b, d, f, g\}$

51. {2, 4, 6}

53. Universal set

55. $\varnothing$

57. $X \cup Y = \{1, 2, 3, 4, 5, 6, 7, 9, 11\}$
    $(X \cup Y) \cap Z = \{2, 5, 6, 11\}$

59. $\overline{W} = \{1, 3, 5, 7, 9, 11\}$
    $\overline{X} = \{2, 4, 6, 8, 10, 12\}$
    $\overline{W} \cap \overline{X} = \varnothing$

61. $A \cap B = B$ since $B \subset A$.

63. $\overline{C} = \{\text{all odd natural numbers}\}$
    $B \cup \overline{C} = \{\text{odd natural numbers or multiples of 9}\}$
    $A \cap (B \cup \overline{C}) = \{x \mid x \text{ is an odd multiple of 3 or an}$
    $\qquad\qquad\qquad\qquad\text{even multiple of 9}\}$
    $\qquad\qquad\qquad = \{3, 9, 15, 18, 21, 27, 33, 36, 39, \ldots\}$

65. (a) None

    (b) Phone

    (c) Television

    (d) Internet

    (e) Phone, television

    (f) Phone, Internet

    (g) Television, Internet

    (h) Phone, television, Internet

67. $2^7 = 128$

69. $2^4 = 16$

71. If every element of set A is also in set B, then A is a *subset* of B.

73. 2^n, which includes $\varnothing$

75. A subset is a set in its own right, hence a *collection* of well-defined objects. An element of a set is just an *individual member* of the set.

77. The union of sets A and B consists of all elements that are not in *at least one* of A and B. The intersection of A and B consists of all elements that are *in both* A and B.

79. The set of all elements used in a particular problem or situation is called a universal set.

81. Answers will vary.

83. Answers will vary.

Exercise Set 2-3

For Exercises 1–8, all solutions have the same first two steps, given here.

Step 1   Draw the set diagram and label each area.

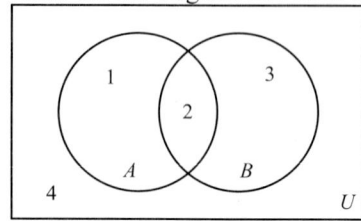

Step 2   From the diagram, list the elements in each set.
$U = \{1, 2, 3, 4\}$
$A = \{1, 2\}$
$B = \{2, 3\}$

1. Step 3   Find the solution to $\overline{A} \cap B$.
$\overline{A} = \{3, 4\}$
$\overline{A} \cap B = \{3\}$

Step 4   Shade area 3.

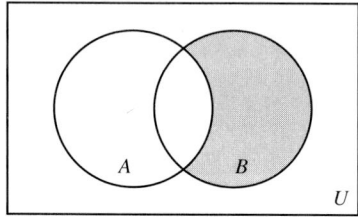

3. Step 3   Find the solution to $\overline{A \cap B}$.
$A \cap B = \{2\}$
$\overline{A \cap B} = \{1, 3, 4\}$

Step 4   Shade areas 1, 3, and 4.

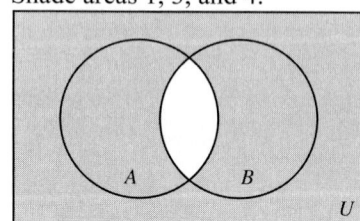

5. Step 3   Find the solution to $\overline{A} \cup \overline{B}$.
$\overline{A} = \{3, 4\}$
$\overline{B} = \{1, 4\}$
$\overline{A} \cup \overline{B} = \{1, 3, 4\}$

Step 4   Shade areas 1, 3, and 4.

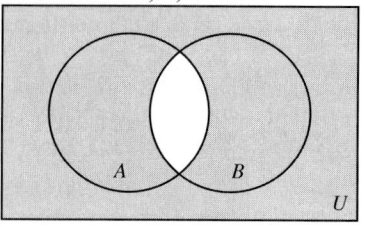

7. Step 3   Find the solution to $\overline{A} \cap \overline{B}$.
$\overline{A} = \{3, 4\}$
$\overline{B} = \{1, 4\}$
$\overline{A} \cap \overline{B} = \{4\}$

Step 4   Shade area 4.

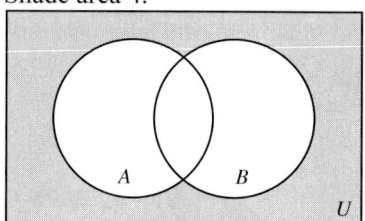

For Exercises 9–26, all solutions have the same first two steps, given here.

Step 1 Draw and label the diagram as shown.

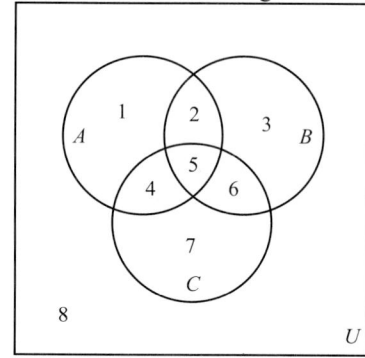

Step 2 From the diagram, list the elements in each set.
$U = \{1, 2, 3, 4, 5, 6, 7, 8\}$
$A = \{1, 2, 4, 5\}$
$B = \{2, 3, 5, 6\}$
$C = \{4, 5, 6, 7\}$

9. Step 3   Find the solution to $A \cup (B \cap C)$.
$B \cap C = \{5, 6\}$
$A \cup (B \cap C) = \{1, 2, 4, 5, 6\}$

Step 4   Shade the sections as shown.

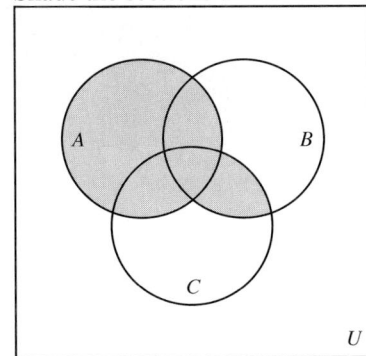

Step 4   Shade the sections as shown.

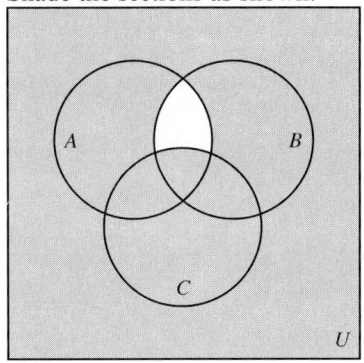

11.   Step 3   Find the solution to $(A \cup B) \cup (A \cap C)$.

$A \cup B = \{1, 2, 3, 4, 5, 6\}$
$A \cap C = \{4, 5\}$
$(A \cup B) \cup (A \cap C) = \{1, 2, 3, 4, 5, 6\}$

Step 4   Shade the sections as shown.

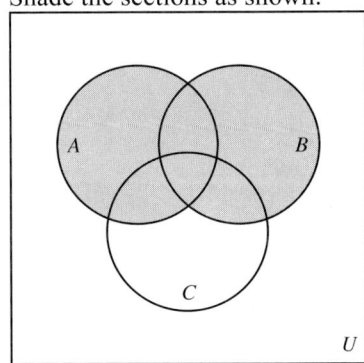

13.   Step 3   Find the solution to $(A \cup B) \cap (A \cup C)$.

$A \cup B = \{1, 2, 3, 4, 5, 6\}$
$A \cup C = \{1, 2, 4, 5, 6, 7\}$
$(A \cup B) \cap (A \cup C) = \{1, 2, 4, 5, 6\}$

Step 4   Shade the sections as shown.

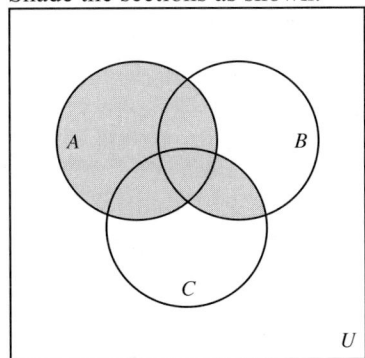

15.   Step 3   Find the solution to $\left(\overline{A \cap B}\right) \cup C$.

$A \cap B = \{2, 5\}$
$\overline{A \cap B} = \{1, 3, 4, 6, 7, 8\}$
$\left(\overline{A \cap B}\right) \cup C = \{1, 3, 4, 5, 6, 7, 8\}$

17.   Step 3   Find the solution to $A \cap \left(\overline{B \cup C}\right)$.

$B \cup C = \{2, 3, 4, 5, 6, 7\}$
$\overline{B \cup C} = \{1, 8\}$
$A \cap \left(\overline{B \cup C}\right) = \{1\}$

Step 4   Shade the sections as shown.

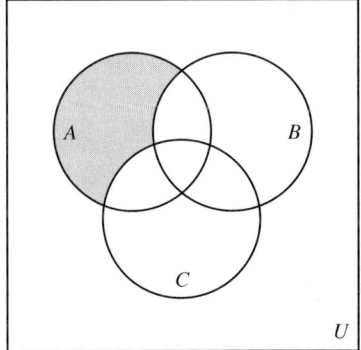

19.   Step 3   Find the solution to $(\overline{A} \cup \overline{B}) \cap C$.

$\overline{A} = \{3, 6, 7, 8\}$
$\overline{B} = \{1, 4, 7, 8\}$
$\overline{A} \cup \overline{B} = \{1, 3, 4, 6, 7, 8\}$
$(\overline{A} \cup \overline{B}) \cap C = \{4, 6, 7\}$

Step 4   Shade the sections as shown.

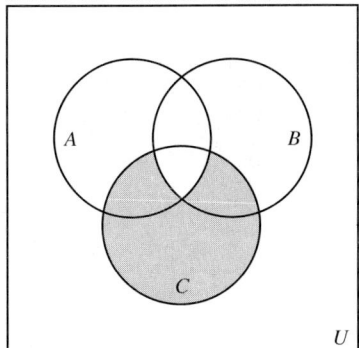

21. Step 3 Find the solution to $\left(\overline{A \cup B}\right) \cap (A \cup C)$.

$A \cup B = \{1, 2, 3, 4, 5, 6\}$
$\overline{A \cup B} = \{7, 8\}$
$A \cup C = \{1, 2, 4, 5, 6, 7\}$
$\left(\overline{A \cup B}\right) \cap (A \cup C) = \{7\}$

Step 4 Shade the section as shown.

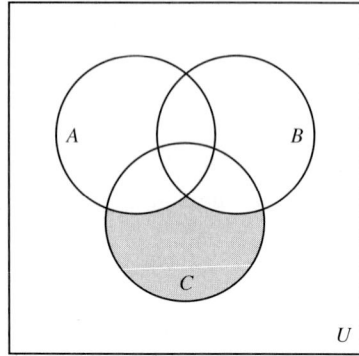

23. Step 3 Find the solution to $\overline{A} \cap \left(\overline{B} \cap \overline{C}\right)$.

$\overline{B} = \{1, 4, 7, 8\}$
$\overline{C} = \{1, 2, 3, 8\}$
$\overline{B} \cap \overline{C} = \{1, 8\}$
$\overline{A} = \{3, 6, 7, 8\}$
$\overline{A} \cap \left(\overline{B} \cap \overline{C}\right) = \{8\}$

Step 4 Shade the section as shown.

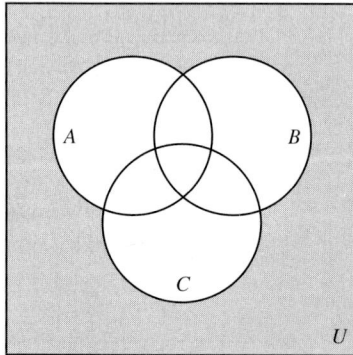

25. Step 3 Find the solution to $\overline{A} \cap \left(\overline{B \cup C}\right)$.

$B \cup C = \{2, 3, 4, 5, 6, 7\}$
$\overline{B \cup C} = \{1, 8\}$
$\overline{A} = \{3, 6, 7, 8\}$
$\overline{A} \cap \left(\overline{B \cup C}\right) = \{8\}$

Step 4 Shade the section as shown.

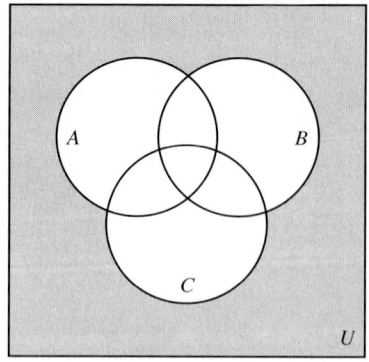

For Exercises 27–33, use the labeled Venn diagram from Step 1 of the solutions for Exercises 9–26.

27. $A \cap B = \{2, 5\}$
$\overline{A \cap B} = \{1, 3, 4, 6, 7, 8\}$
$\overline{A} = \{3, 6, 7, 8\}$
$\overline{B} = \{1, 4, 7, 8\}$
$\overline{A} \cup \overline{B} = \{1, 3, 4, 6, 7, 8\}$
Yes, $\overline{A \cap B}$ is equivalent to $\overline{A} \cup \overline{B}$.

29. $(A \cup B) \cup C = \{1, 2, 3, 4, 5, 6, 7\}$
$A \cup (B \cup C) = \{1, 2, 3, 4, 5, 6, 7\}$
Yes, $(A \cup B) \cup C$ is equivalent to $A \cup (B \cup C)$.

31. $\overline{C} = \{1, 2, 3, 8\}$
$B \cap \overline{C} = \{2, 3\}$
$\overline{A} = \{3, 6, 7, 8\}$
$\overline{A} \cup (B \cap \overline{C}) = \{2, 3, 6, 7, 8\}$
$\overline{A} \cup B = \{2, 3, 5, 6, 7, 8\}$
$(\overline{A} \cup B) \cap \overline{C} = \{2, 3, 8\}$
No, $\overline{A} \cup (B \cap \overline{C})$ is not equivalent to $(\overline{A} \cup B) \cap \overline{C}$.

33. $A \cap B = \{2, 5\}$
$\overline{A \cap B} = \{1, 3, 4, 6, 7, 8\}$
$\left(\overline{A \cap B}\right) \cup C = \{1, 3, 4, 5, 6, 7, 8\}$
$\overline{A} = \{3, 6, 7, 8\}$
$\overline{B} = \{1, 4, 7, 8\}$
$\overline{A} \cup \overline{B} = \{1, 3, 4, 6, 7, 8\}$
$(\overline{A} \cup \overline{B}) \cap C = \{4, 6, 7\}$
No, $\left(\overline{A \cap B}\right) \cup C$ is not equivalent to $(\overline{A} \cup \overline{B}) \cap C$.

35. In the figure are shown all those elements (represented by points) that belong: (1) strictly to A, (2) strictly to B, and (3) to both A and B. These mentioned points form precisely the union of A and B.

37. Disjoint sets have no elements in common; accordingly, the circles representing these sets cannot touch each other anywhere.

39. $B - C = B \cap \overline{C}$

$\overline{C} = \{1, 2, 3, 8\}$

$B \cap \overline{C} = \{2, 3\}$, so $B - C = \{2, 3\}$.

41. $A - C = A \cap \overline{C}$

$\overline{C} = \{1, 2, 3, 8\}$

$A \cap \overline{C} = \{1, 2\}$, so $A - C = \{1, 2\}$.

43. $A - B = A \cap \overline{B}$

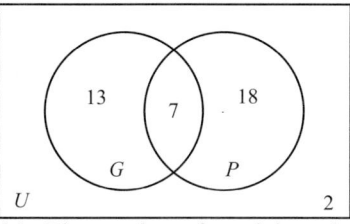

Exercise Set 2-4

1. Step 1 Draw a Venn diagram, where
U = universal set
G = number of people who played guitar
P = number of people who played piano

Step 2 Since 7 people played both instruments, place 7 in the intersection.

Step 3 Since 20 people played guitar and 7 people played both instruments, subtract $20 - 7 = 13$ to get the number of people who played guitar only. By subtracting, find the number of people who played piano only; $25 - 7 = 18$. Place these numbers in the appropriate section of the Venn diagram.

Step 4 Find the number of people who played neither instrument by adding, $13 + 7 + 18 = 38$, and subtracting that number from the total number of musicians, 40; $40 - 38 = 2$. Place 2 in the appropriate section of the Venn diagram.

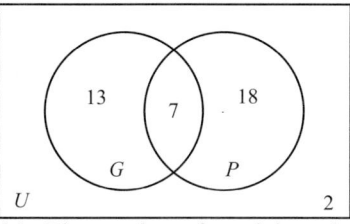

(a) The number of musicians that played guitar only is 13.
(b) The number of musicians that played piano only is 18.
(c) The number of musicians that played neither instrument is 2.

3. Step 1 Draw a Venn diagram, where
U = universal set
R = number of red automobiles
W = number of white automobiles

Step 2 Since 3 automobiles were two-toned, place 3 in the intersection.

Step 3 Since 8 automobiles were red and 3 automobiles were two-toned, subtract $8 - 3 = 5$ to get the number of automobiles that were red only. By subtracting, find the number of automobiles that were white only; $5 - 3 = 2$. Place these numbers in the appropriate section of the Venn diagram.

Step 4 Find the number of automobiles that were not red or white by adding, $3 + 5 + 2 = 10$, and subtracting that number from the total number of automobiles; $16 - 10 = 6$. Place 6 in the appropriate section of the Venn diagram.

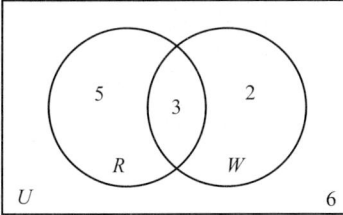

(a) The number of automobiles that were red only is 5.
(b) The number of automobiles that were red or white but not two-toned is $5 + 2 = 7$.
(c) The number of automobiles that were neither red nor white is 6.

5. Step 1 Draw a Venn diagram.

Step 2 Since 2 women participated in all three activities, place 2 where all three sets intersect.

Step 3 Find the number of women that played golf and fished but did not play racquetball. Subtract the number of women that participate in all three activities (2) from the number of women that played golf and fished (9), $9 - 2 = 7$.
Find the number of women that fished and played racquetball but did not play golf by subtracting $11 - 2 = 9$.
Find the number of women that played golf and racquetball but did not fish; $7 - 2 = 5$.
Place these numbers in the appropriate section of the Venn diagram.

Step 4 By subtracting, find the number of women that played only golf; $16 - (7 + 5 + 2) = 2$.
By subtracting, find the number of women that fished only; $24 - (7 + 9 + 2) = 6$.
By subtracting, find the number of women that played only racquetball; $20 - (9 + 5 + 2) = 4$.
Place these numbers in the appropriate section of the Venn diagram.

Step 5 Find the number of women that did not participate in any of the activities by adding all the numbers, $2 + 7 + 9 + 5 + 2 + 6 + 4 = 35$, and subtracting that number from the total number of women, 70; $70 - 35 = 35$.
Place 35 in the appropriate section of the Venn diagram.

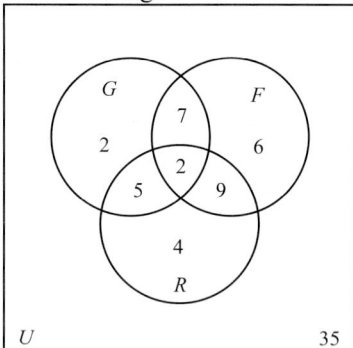

(a) The number of women that played only golf is 2.
(b) The number of women that fished and played racquetball but not golf is 9.
(c) The number of women that did not participate in any of the three activities is 35.

7. Step 1 Draw a Venn diagram.

Step 2 Since 2 cars failed because of all three conditions, place 2 where all three sets intersect.

Step 3 Find the number of cars that failed because of bad tires and brakes but not exhaust systems. Subtract the number of cars that failed because of all three conditions (2) from the number of cars that failed because of bad tires and brakes (9), $9 - 2 = 7$.
Find the number of cars that failed because of bad brakes and exhaust systems but not tires by subtracting $8 - 2 = 6$.
Find the number of cars that failed because of bad tires and exhaust systems but not brakes by subtracting $5 - 2 = 3$.
Place these numbers in the appropriate section of the Venn diagram.

Step 4 By subtracting, find the number of cars that failed because of bad tires only; $14 - (2 + 7 + 3) = 2$.
By subtracting, find the number of cars that failed because of bad brakes only; $23 - (2 + 7 + 6) = 8$.
By subtracting, find the number of cars that failed because of faulty exhaust systems; $15 - (2 + 6 + 3) = 4$.
Place these numbers in the appropriate section of the Venn diagram.

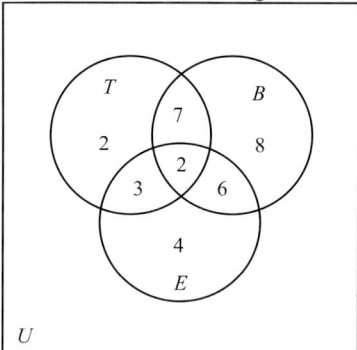

(a) The number of cars that failed because of at least two of these problems is $7 + 3 + 6 + 2 = 18$.
(b) The number of cars that failed because of tires and brakes but not exhaust systems is 7.
(c) The number of cars that failed because of only one item is $2 + 8 + 4 = 14$.
(d) The number of cars that failed because of tires or brakes but not exhaust systems is $2 + 7 + 8 = 17$.

9. Step 1 Draw a Venn diagram.

Step 2   Since 3 vendors sold all three items, place 3 where all three sets intersect.

Step 3   Find the number of vendors who sold cards and picture frames but not sunglasses. Subtract the number of vendors who sold all three items (3) from the number of vendors who sold cards and picture frames (9), $9 - 3 = 6$. Find the number of vendors who sold picture frames and sunglasses but not cards by subtracting $8 - 3 = 5$. Find the number of vendors who sold cards and sunglasses but not picture frames by subtracting $5 - 3 = 2$. Place these numbers in the appropriate section of the Venn diagram.

Step 4   By subtracting, find the number of vendors who sold cards only; $19 - (3 + 6 + 2) = 8$. By subtracting, find the number of vendors who sold picture frames only; $21 - (3 + 6 + 5) = 7$. By subtracting, find the number of vendors who sold sunglasses only; $19 - (3 + 5 + 2) = 9$. Place these numbers in the appropriate section of the Venn diagram.

Step 5   Find the number of vendors who sold none of these items by adding all the numbers, $3 + 6 + 5 + 2 + 8 + 7 + 9 = 40$, and subtracting that number from the total number of vendors, 70; $70 - 40 = 30$. Place 30 in the appropriate section of the Venn diagram.

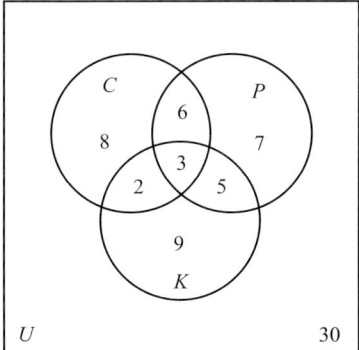

(a)   The number of vendors who sold at most two of these items is $8 + 6 + 7 + 5 + 9 + 2 = 37$.

(b)   The number of vendors who sold cards and sunglasses but not picture frames is 2.

(c)   The number of vendors who sold neither cards nor sunglasses is $7 + 30 = 37$.

11. Step 1   Draw a Venn diagram.

Step 2   Since 3 briefcases contained all three items, place 3 where all three sets intersect.

Step 3   Find the number of briefcases that contained paper clips and pencils but not pens. Subtract the number of briefcases that contained all three items (3) from the number of briefcases that contained paper clips and pencils (9), $9 - 3 = 6$. Find the number of briefcases that contained pencils and pens but not paper clips by subtracting $15 - 3 = 12$. Find the number of briefcases that contained paper clips and pens but not pencils by subtracting $6 - 3 = 3$. Place these numbers in the appropriate section of the Venn diagram.

Step 4   By subtracting, find the number of briefcases that contained paper clips only; $20 - (3 + 6 + 3) = 8$. By subtracting, find the number of briefcases that contained pencils only; $26 - (3 + 6 + 12) = 5$. By subtracting, find the number of briefcases that contained pens only; $20 - (3 + 12 + 3) = 2$. Place these numbers in the appropriate section of the Venn diagram.

Step 5   Find the number of briefcases that contained none of these items by adding all the number, $3 + 6 + 12 + 3 + 8 + 5 + 2 = 39$, and subtracting that number from the total number of briefcases, 61; $61 - 39 = 22$. Place 22 in the appropriate section of the Venn diagram.

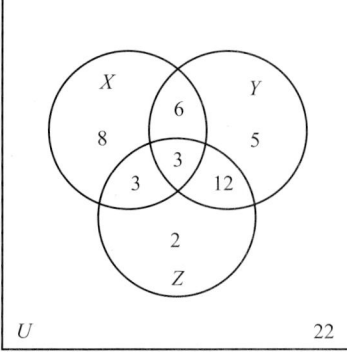

X represents paper clips
Y represents pencils
Z represents pens

(a)   The number of briefcases that contained none of the three articles is 22.

(b) The number of briefcases that contained exactly two of the three items is $6 + 12 + 3 = 21$.

(c) The number of briefcases that contained exactly one of the three items is $8 + 5 + 2 = 15$.

13. **Step 1**   Draw a Venn diagram.

Step 2   Since 10 people participated in all three activities, place 10 where all three sets intersect.

Step 3   Find the number of people who visited the casinos and went shopping but did not go swimming. Subtract the number of people who participated in all three activities (10) from the number of people who went shopping and visited the casinos (25), $25 - 10 = 15$. Find the number of people who went shopping and swimming but did not visit the casinos by subtracting $14 - 10 = 4$. Find the number of people who went swimming and to the casinos but did not go shopping by subtracting $18 - 10 = 8$. Place these numbers in the appropriate section of the Venn diagram.

Step 4   By subtracting, find the number of people who visited the casinos only; $39 - (15 + 10 + 8) = 6$. By subtracting, find the number of people who went shopping only; $51 - (15 + 10 + 4) = 22$. By subtracting, find the number of people who went swimming only; $26 - (8 + 10 + 4) = 4$. Place these numbers in the appropriate section of the Venn diagram.

Step 5   Find the number of people who did not participate in any of these activities by adding all the numbers, $6 + 15 + 22 + 8 + 10 + 4 + 4 = 69$, and subtracting that number from the total number of visitors, 121; $121 - 69 = 52$. Place 52 in the appropriate section of the Venn diagram.

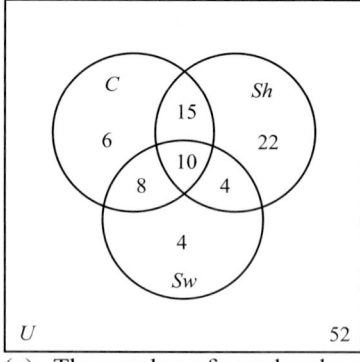

(a) The number of people who did exactly one thing is $6 + 22 + 4 = 32$.

(b) The number of people who went shopping but not swimming is $15 + 22 = 37$.

(c) The number of people who went neither shopping nor swimming is $6 + 52 = 58$.

15. The number of growers who grew either apples or pears is $34 - 2 = 32$.
The number of growers who grew apples but not pears is $32 - 20 = 12$.
The number of growers who grew pears but not apples is $32 - 18 = 14$.
The number of growers who grew both pears and apples is $32 - (12 + 14) = 6$.

Exercise Set 2-5

1. Use inductive reasoning.
$7(1) = 7$
$7(2) = 14$
$7(3) = 21$
$7(4) = 28$
$7(5) = 35$
etc.
The general term is $7n$.

3. Use inductive reasoning.
$4^1 = 4$
$4^2 = 16$
$4^3 = 64$
$4^4 = 256$
$4^5 = 1024$
etc.
The general term is 4^n.

5. Use inductive reasoning.
$-3(1) = -3$
$-3(2) = -6$
$-3(3) = -9$
$-3(4) = -12$
$-3(5) = -15$
etc.
The general term is $-3n$.

7. Use inductive reasoning.
$$\frac{1}{1+1} = \frac{1}{2}$$
$$\frac{1}{2+1} = \frac{1}{3}$$
$$\frac{1}{3+1} = \frac{1}{4}$$
$$\frac{1}{4+1} = \frac{1}{5}$$
$$\frac{1}{5+1} = \frac{1}{6}$$
etc.

The general term is $\dfrac{1}{n+1}$.

9. Use inductive reasoning.
$4(1) - 2 = 2$
$4(2) - 2 = 6$
$4(3) - 2 = 10$
$4(4) - 2 = 14$
$4(5) - 2 = 18$
etc.
The general term is $4n - 2$.

11. $\{3, \ 6, \ 9, \ 12, \ 15, \ ..., \ 3n, \ ...\}$
    $\updownarrow \ \updownarrow \ \updownarrow \ \updownarrow \ \updownarrow \qquad \updownarrow$
$\{6, 12, 18, 24, 30, \ ..., \ 6n, \ ...\}$

13. $\{9, \ 18, \ 27, \ 36, \ 45, \ ..., \ 9n, \ ...\}$
    $\updownarrow \ \updownarrow \ \updownarrow \ \updownarrow \ \updownarrow \qquad \updownarrow$
$\{18, 36, 108, 144, 180, \ ..., \ 18n, \ ...\}$

15. $\{2, \ 5, \ 8, \ 11, \ ..., \ 3n-1, \ ...\}$
    $\updownarrow \ \updownarrow \ \updownarrow \ \updownarrow \qquad \updownarrow$
$\{5, 11, 17, 23, \ ..., \ 6n-1, \ ...\}$

17. $\{10, \quad 100, \ ..., \ 10^n, \ ...\}$
    $\updownarrow \qquad \updownarrow \qquad \updownarrow$
$\{100, 10,000, \ ..., \ 10^{2n}, \ ...\}$

19. $\left\{\dfrac{5}{1}, \dfrac{5}{2}, \dfrac{5}{3}, \ ..., \ \dfrac{5}{n}, \ ...\right\}$
    $\updownarrow \ \updownarrow \ \updownarrow \qquad \updownarrow$
$\left\{\dfrac{5}{2}, \dfrac{5}{3}, \dfrac{5}{4}, \ ..., \ \dfrac{5}{n+1}, \ ...\right\}$

21. An infinite set is a set that can be placed in a one-to-one correspondence with a proper subset of itself.

23. A set is countable if it is finite or if there is a one-to-one correspondence between the members of the set and the natural numbers.

25. $\aleph_0$

Review Exercises

1. $\{52, 54, 56, 58\}$

3. $\{l, e, t, r\}$

5. $\{501, 502, 503, ...\}$

7. $\varnothing$

9. $\{x \mid x \in E \text{ and } 16 < x < 26\}$

11. $\{x \mid x \text{ is an odd natural number greater than } 100\}$

13. Infinite

15. Finite

17. Finite

19. $\varnothing$; $\{r\}$; $\{s\}$; $\{t\}$; $\{r, s\}$; $\{r, t\}$; $\{s, t\}$; $\{r, s, t\}$

21. $2^5 = 32$ subsets; 31 proper subsets

23. $\{t, u, v\}$

25. $A \cap B = \{t, u, v\}$
$(A \cap B) \cap C = \varnothing$

27. $A \cup B = \{p, r, t, u, v, x, y\}$
$\overline{A \cup B} = \{q, s, w, z\}$
$\left(\overline{A \cup B}\right) \cap C = \{s, w, z\}$

29. $B \cup C = \{s, t, u, v, w, x, y, z\}$
$\overline{A} = \{q, s, w, x, y, z\}$
$(B \cup C) \cap \overline{A} = \{s, w, x, y, z\}$

31. $\overline{B} = \{p, q, r, s, w, z\}$
$\overline{C} = \{p, q, r, t, u, v, x, y\}$
$\overline{B} \cap \overline{C} = \{p, q, r\}$
$\overline{A} = \{q, s, w, x, y, z\}$
$(\overline{B} \cap \overline{C}) \cup \overline{A} = \{p, q, r, s, w, x, y, z\}$

33. Step 1   Draw the set diagram and label each area.

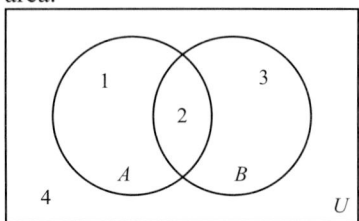

Step 2   From the diagram, list the elements in each set.
$U = \{1, 2, 3, 4\}$
$A = \{1, 2\}$
$B = \{2, 3\}$

Step 3   Find the solution to $\overline{A} \cap B$.
$\overline{A} = \{3, 4\}$
$\overline{A} \cap B = \{3\}$

Step 4   Shade area 3.

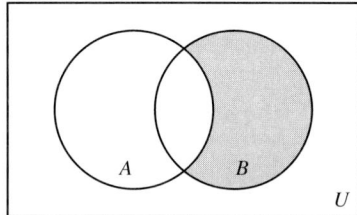

35. Step 1   Draw and label the diagram as shown.

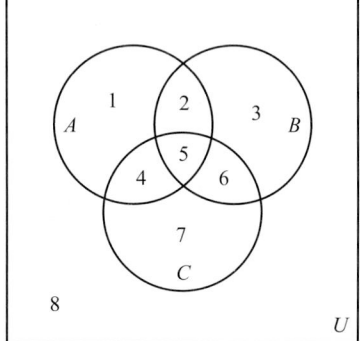

Step 2   From the diagram, list the elements in each set.
$U = \{1, 2, 3, 4, 5, 6, 7, 8\}$
$A = \{1, 2, 4, 5\}$
$B = \{2, 3, 5, 6\}$
$C = \{4, 5, 6, 7\}$

Step 3   Find the solution to $(\overline{A} \cap \overline{B}) \cup C$.
$\overline{A} = \{3, 6, 7, 8\}$
$\overline{B} = \{1, 4, 7, 8\}$
$\overline{A} \cap \overline{B} = \{7, 8\}$
$(\overline{A} \cap \overline{B}) \cup C = \{4, 5, 6, 7, 8\}$

Step 4   Shade the sections as shown.

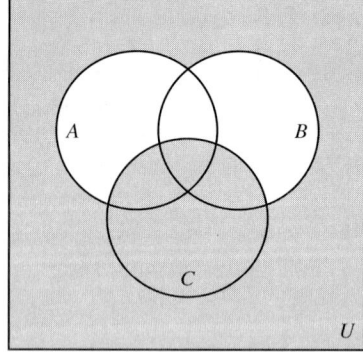

37. Step 1   Draw a Venn diagram.

Step 2   Since 2 students had both cereal and toast, place 2 where the sets intersect.

Step 3   Since 10 students had cereal for breakfast and 2 had both cereal and toast, subtract $10 - 2 = 8$ to get the number of students who had cereal only. By subtracting, find the number of students who had toast only; $5 - 2 = 3$.

Step 4   Find the number of students who had neither cereal nor toast by adding $2 + 8 + 3 = 13$, and subtracting that number from the total number of students, 25; $25 - 13 = 12$.

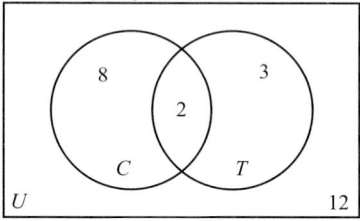

(a)   The number of students who did not eat cereal or toast is 12.
(b)   The number of students who had toast only is 3.

39. Step 1   Draw a Venn diagram.

Step 2   Since 6 students listened to all three, place 6 where all three sets intersect.

Step 3   Find the number of students who listened to the radio and tapes but not CDs by subtracting $8 - 6 = 2$.
Find the number of students who listened to tapes and CDs but not the radio by subtracting $13 - 6 = 7$.
Find the number of students who listened to the radio and CDs but not tapes by subtracting $11 - 6 = 5$.

Step 4 Find the number of students who listened to only the radio by subtracting $22 - (6 + 2 + 5) = 9$.
Find the number of students who listened only to tapes by subtracting $18 - (6 + 2 + 7) = 3$.
Find the number of students who listened only to CDs by subtracting $33 - (6 + 7 + 5) = 15$.

Step 5 Find the number of students who didn't listen to any of these three types of music by adding,
$6 + 2 + 7 + 5 + 9 + 3 + 15 = 47$, and subtracting that number from the total number of students, 53; $53 - 47 = 6$.

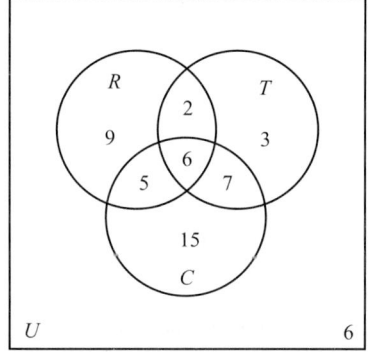

(a) The number of students who listened to tapes only is 3.
(b) The number of students who listened to the radio and CDs but not tapes is 5.
(c) The number of students who did not listen to any of the three types of music is 6.

41. False; the elements of the sets are not the same.

43. False; 100 is an element of {80, 100, 120, ...} but not an element of {40, 80, 120, ...}.

45. True

47. False; {5, 7} is a proper subset of {5, 6, 7}.

49. True

51. False; for any set $A \cap \varnothing = \varnothing$.

53. Use inductive reasoning.
$-3 - 2(1) = -5$
$-3 - 2(2) = -7$
$-3 - 2(3) = -9$
$-3 - 2(4) = -12$
$-3 - 2(5) = -15$
etc.
The general term is $-3 - 2n$.

Chapter Test

1. {92, 94, 96, 98}

3. {e, n, v, l, o, p}

5. {1, 2, 3, 4, ..., 79}

7. {January, June, July}

9. $\{x \mid x \in E \text{ and } 10 < x < 20\}$

11. $\{x \mid x \text{ is an odd natural number greater than 200}\}$

13. Infinite

15. Finite

17. Finite

19. { } or $\varnothing$; {p}; {q}; {r}; {p, q}; {p, r}; {q, r}; {p, q, r}

21. {a}

23. {b, c, d, e, f, h}

25. $\overline{B} = \{b, c, d, e, f, h\}$
$A \cap \overline{B} = \{b, d, e, f\}$
$\overline{C} = \{a, b, c, d, f, g, i, k\}$
$(A \cap \overline{B}) \cup \overline{C} = \{a, b, c, d, e, f, g, i, k\}$

27. Step 1 Draw the set diagram and label each area.

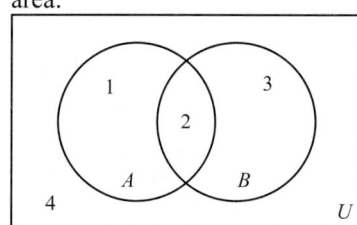

Step 2 From the diagram, list the elements in each set.
$U = \{1, 2, 3, 4\}$
$A = \{1, 2\}$
$B = \{2, 3\}$

Step 3 Find the solution to $\overline{A \cap B}$.
$A \cap B = \{2\}$
$\overline{A \cap B} = \{1, 3, 4\}$

Step 4   Shade areas 1, 3, and 4.

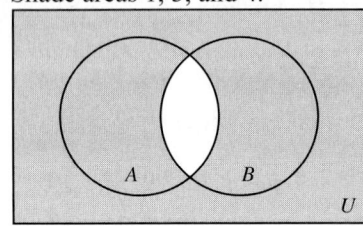

29. Step 1   Draw and label the diagram as shown.

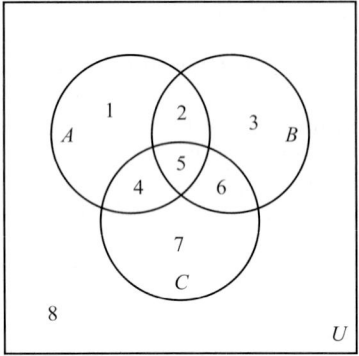

Step 2   From the diagram, list the elements in each set.
$U = \{1, 2, 3, 4, 5, 6, 7, 8\}$
$A = \{1, 2, 4, 5\}$
$B = \{2, 3, 5, 6\}$
$C = \{4, 5, 6, 7\}$

Step 3   Find the solution to $A \cap \left(\overline{B \cup C} \right)$.

$B \cup C = \{2, 3, 4, 5, 6, 7\}$
$\overline{B \cup C} = \{1, 8\}$
$A \cap \left(\overline{B \cup C} \right) = \{1\}$

Step 4   Shade the sections as shown.

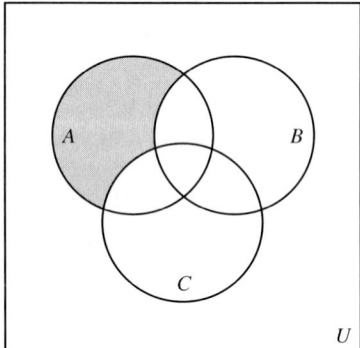

31.   Use inductive reasoning.
$15(1) = 15$
$15(2) = 30$
$15(3) = 45$
$15(4) = 60$
$15(5) = 75$
The general term is $15n$.

3 THE ART OF PROBLEM SOLVING

Exercise Set 3-1

1. No

3. Yes

5. No

7. Yes

9. No

11. Compound

13. Compound

15. Simple

17. Compound

19. Simple

21. The sky is not blue.

23. The blanket is red.

25. It is true that Harry failed statistics.

27. Conjunction

29. Biconditional

31. Disjunction

33. Biconditional

35. $p \wedge q$

37. $(\sim q) \to p$

39. $\sim (\sim q)$

41. $q \vee \sim p$

43. $q \leftrightarrow p$

45. $\sim q$

47. $q \to p$

49. $(\sim q) \vee p$

51. $q \leftrightarrow p$

53. $(\sim q) \to p$

55. The plane is on time, and the sky is clear.

57. If the sky is clear, then the plane is on time.

59. The plane is not on time, and the sky is not clear.

61. The plane is on time, or the sky is not clear.

63. If the sky is clear, then the plane is or is not on time.

65. Trudy is not attractive.

67. Mark is handsome, or Trudy is not attractive.

69. If Mark is not handsome, then Trudy is not attractive.

71. Mark is handsome, or Trudy is attractive.

73. Trudy is attractive, or Mark is handsome.

75. A statement is a declarative sentence that can be classified as true or false, but not both.

77. $\wedge$; $\vee$; $\to$; $\leftrightarrow$

79. It cannot be classified as true or false.

Exercise Set 3-2

1.

p	q	$\sim$	$(p \vee q)$
T	T	F	T
T	F	F	T
F	T	F	T
F	F	T	F
		②	①

3.

p	q	$\sim p$	$\wedge$	q
T	T	F	F	T
T	F	F	F	F
F	T	T	T	T
F	F	T	F	F
		①	③	②

5.

p	q	$\sim p$	$\leftrightarrow$	q
T	T	F	F	T
T	F	F	T	F
F	T	T	T	T
F	F	T	F	F
		①	③	②

7.

p	q	$\sim$	$(p \wedge q)$	$\rightarrow$	p
T	T	F	T	T	T
T	F	T	F	T	T
F	T	T	F	F	F
F	F	T	F	F	F
		②	①	④	③

9.

p	q	$(\sim q$	$\wedge$	$p)$	$\rightarrow$	$\sim p$
T	T	F	F	T	T	F
T	F	T	T	T	F	F
F	T	F	F	F	T	T
F	F	T	F	F	T	T
		①	③	②	⑤	④

11.

p	q	$(p \wedge q)$	$\leftrightarrow$	$(q$	$\vee$	$\sim p)$
T	T	T	T	T	T	F
T	F	F	T	F	F	F
F	T	F	F	T	T	T
F	F	F	F	F	T	T
		①	⑤	③	④	②

13.

p	q	$(p \wedge q)$	$\vee$	p
T	T	T	T	T
T	F	F	T	T
F	T	F	F	F
F	F	F	F	F
		①	③	②

15.

p	q	r	(r ∧ q)	∨	(p ∧ q)
T	T	T	T	T	T
T	T	F	F	T	T
T	F	T	F	F	F
T	F	F	F	F	F
F	T	T	T	T	F
F	T	F	F	F	F
F	F	T	F	F	F
F	F	F	F	F	F
			①	③	②

17.

p	q	r	~	(p ∨ q)	→	~	(p ∧ r)
T	T	T	F	T	T	F	T
T	T	F	F	T	T	T	F
T	F	T	F	T	T	F	T
T	F	F	F	T	T	T	F
F	T	T	F	T	T	T	F
F	T	F	F	T	T	T	F
F	F	T	T	F	T	T	F
F	F	F	T	F	T	T	F
			②	①	⑤	④	③

19.

p	q	r	(~p	∨	q)	∧	r
T	T	T	F	T	T	T	T
T	T	F	F	T	T	F	F
T	F	T	F	F	F	F	T
T	F	F	F	F	F	F	F
F	T	T	T	T	T	T	T
F	T	F	T	T	T	F	F
F	F	T	T	T	F	T	T
F	F	F	T	T	F	F	F
			①	③	②	⑥	⑤

21.

p	q	r	$(p \wedge q)$	$\leftrightarrow$	$(\sim r$	$\vee$	$q)$
T	T	T	T	T	F	T	T
T	T	F	T	T	T	T	T
T	F	T	F	T	F	F	F
T	F	F	F	F	T	T	F
F	T	T	F	F	F	T	T
F	T	F	F	F	T	T	T
F	F	T	F	T	F	F	F
F	F	F	F	F	T	T	F
			①	⑤	②	④	③

23.

p	q	r	r	$\rightarrow$	$\sim$	$(p \vee q)$
T	T	T		F	F	T
T	T	F		T	F	T
T	F	T		F	F	T
T	F	F		T	F	T
F	T	T		F	F	T
F	T	F		T	F	T
F	F	T		T	T	F
F	F	F		T	T	F
				③	②	①

25.

p	q	r	p	$\rightarrow$	$(\sim q$	$\wedge$	$\sim r)$
T	T	T	T	F	F	F	F
T	T	F	T	F	F	F	T
T	F	T	T	F	T	F	F
T	F	F	T	T	T	T	T
F	T	T	F	T	F	F	F
F	T	F	F	T	F	F	T
F	F	T	F	T	T	F	F
F	F	F	F	T	T	T	T
			①	⑤	②	④	③

27.

p	q	r	~	(q → p)	∧	r
T	T	T	F	T	F	T
T	T	F	F	T	F	F
T	F	T	F	T	F	T
T	F	F	F	T	F	F
F	T	T	T	F	T	T
F	T	F	T	F	F	F
F	F	T	F	T	F	T
F	F	F	F	T	F	F
			②	①	④	③

29.

p	q	r	(r ∨ q)	∧	(r ∧ p)
T	T	T	T	T	T
T	T	F	T	F	F
T	F	T	T	T	T
T	F	F	F	F	F
F	T	T	T	F	F
F	T	F	T	F	F
F	F	T	T	F	F
F	F	F	F	F	F
			①	③	②

31. A truth table enables us to decide on the truth value of a statement (usually a compound one) by examining the truth value of the components of the statement.

33. The *inclusive* disjunction is false only when both of its parts are false. The *exclusive* disjunction is false not only when both parts are false but also when both parts are true. At other times it's true.

35. ↔; →; ∧ or ∨; ~

Exercise Set 3-3

1.

p	q	(p ∨ q)	∨	(~p	∧	~q)
T	T	T	T	F	F	F
T	F	T	T	F	F	T
F	T	T	T	T	F	F
F	F	F	T	T	T	T
			↑			

Since the truth table value consists of all Ts, the statement is a tautology.

3.

p	q	(p ∧ q)	∧	(~p	∨	~q)
T	T	T	F	F	F	F
T	F	F	F	F	T	T
F	T	F	F	T	T	F
F	F	F	F	T	T	T
			↑			

Since the truth table value consists of all Fs, the statement is a self-contradiction.

5.

p	q	(p ↔ q)	∨	~	(q ↔ p)
T	T	T	T	F	T
T	F	F	T	T	F
F	T	F	T	T	F
F	F	T	T	F	T
			↑		

Since the truth table value consists of all Ts, the statement is a tautology.

7.

p	q	(p ∨ q)	∧	(~p	∨	~q)
T	T	T	F	F	F	F
T	F	T	T	F	T	T
F	T	T	T	T	T	F
F	F	F	F	T	T	T
			↑			

Since the statement can be true in some cases and false in some cases, it is neither a tautology nor a self-contradiction.

9.

p	q	(p ↔ q)	∧	(~p	↔	~q)
T	T	T	T	F	T	F
T	F	F	F	F	F	T
F	T	F	F	T	F	F
F	F	T	T	T	T	T
			↑			

Since the statement can be true in some cases and false in some cases, it is neither a tautology nor a self-contradiction.

11.

p	q	~q	→	p	~p	→	q
T	T	F	T	T	F	T	T
T	F	T	T	T	F	T	F
F	T	F	T	F	T	T	T
F	F	T	F	F	T	F	F
			↑			↑	

Since both statements have the same truth values, they are logically equivalent.

13.

p	q	~	(p ∨ q)	p	→	~q
T	T	F	T	T	F	F
T	F	F	T	T	T	T
F	T	F	T	F	T	F
F	F	T	F	F	T	T
		↑			↑	

The statements are neither logically equivalent nor negations.

15.

p	q	q → p	~	(p → q)
T	T	T	F	T
T	F	T	T	F
F	T	F	F	T
F	F	T	F	T
		↑	↑	

The statements are neither logically equivalent nor negations.

17.

p	q	~	(p ∨ q)	~	(~p	∧	~q)
T	T	F	T	T	F	F	F
T	F	F	T	T	F	F	T
F	T	F	T	T	T	F	F
F	F	T	F	F	T	T	T
		↑		↑			

Since the truth values of the two statements are exactly opposites, the statements are negations.

19.

p	q	r	(p ∧ q)	∨	r	p	∧	(q ∨ r)
T	T	T	T	T	T	T	T	T
T	T	F	T	T	F	T	T	T
T	F	T	F	T	T	T	T	T
T	F	F	F	F	F	T	F	F
F	T	T	F	T	T	F	F	T
F	T	F	F	F	F	F	F	T
F	F	T	F	T	T	F	F	T
F	F	F	F	F	F	F	F	F
				↑			↑	

The statements are neither logically equivalent nor negations.

21. $q \to p$; $\sim p \to \sim q$; $\sim q \to \sim p$

23. $p \to \sim q$; $q \to \sim p$; $\sim p \to q$

25. $\sim q \to p$; $\sim p \to q$; $q \to \sim p$

27. *Converse*: If he will get a job, then he graduated.
Inverse: If he did not graduate, then he will not get a job.
Contrapositive: If he will not get a job, then he did not graduate.

29. *Converse*: If I have a party, then it is my birthday.
Inverse: If it is not my birthday, then I will not have a party.
Contrapositive: If I will not have a party, then it is not my birthday.

31. When, in their respective truth tables, they have the same value (T or F) in the same relative position.

33. The converse, the inverse, and the contrapositive can be made from the conditional statement.

35. $p \land \sim q$

37. Answers will vary.

Exercise Set 3-4

1.

p	q	(p → q)	∧	(p ∧ q)	⇒	p
T	T	T	T	T	T	T
T	F	F	F	F	T	T
F	T	T	F	F	T	F
F	F	T	F	F	T	F

The argument is valid.

3.

p	q	(p ∨ q)	∧	~q	⇒	p
T	T	T	F	F	T	T
T	F	T	T	T	T	T
F	T	T	F	F	T	F
F	F	F	F	T	T	F

The argument is valid.

5.

p	q	$(p$	$\leftrightarrow$	$\sim q)$	$\wedge$	$(p$	$\wedge$	$\sim q)$	$\Rightarrow$	$p \vee q$
T	T	T	F	F	F	T	F	F	T	T
T	F	T	T	T	T	T	T	T	T	T
F	T	F	T	F	F	F	F	F	T	T
F	F	F	F	T	F	F	F	T	T	F

The argument is valid.

7.

p	q	$(p \vee q)$	$\wedge$	$(\sim p$	$\wedge$	$\sim q)$	$\Rightarrow$	p
T	T	T	F	F	F	F	T	T
T	F	T	F	F	F	T	T	T
F	T	T	F	T	F	F	T	F
F	F	F	F	T	T	T	T	F

The argument is valid.

9.

p	q	$(p$	$\vee$	$\sim q)$	$\wedge$	$(\sim q$	$\rightarrow$	$p)$	$\Rightarrow$	p
T	T	T	T	F	T	F	T	T	T	T
T	F	T	T	T	T	T	T	T	T	T
F	T	F	F	F	F	F	T	F	T	F
F	F	F	T	T	F	T	F	F	T	F

The argument is valid.

11.

p	q	r	$(p$	$\wedge$	$\sim q)$	$\wedge$	$(\sim r$	$\rightarrow$	$q)$	$\Rightarrow$	q
T	T	T	T	F	F	F	F	T	T	T	T
T	T	F	T	F	F	F	T	T	T	T	T
T	F	T	T	T	T	T	F	T	F	F	F
T	F	F	T	T	T	F	T	F	F	T	F
F	T	T	F	F	F	F	F	T	T	T	T
F	T	F	F	F	F	F	T	T	T	T	T
F	F	T	F	F	T	F	F	T	F	T	F
F	F	F	F	F	T	F	T	F	F	T	F

The argument is not valid.

13. Let p = "It rains." Let q = "I (will) do my homework."

$p \rightarrow q$

$\dfrac{\sim p}{\therefore \sim q}$

p	q	$(p \rightarrow q)$	$\wedge$	$\sim p$	$\Rightarrow$	$\sim q$
T	T	T	F	F	T	F
T	F	F	F	F	T	T
F	T	T	T	T	F	F
F	F	T	T	T	T	T

The argument is invalid.

15. Let p = "Sam gains 15 pounds." Let q = "He makes the football team."

$p \rightarrow q$

$\dfrac{q}{\therefore p}$

p	q	$(p \rightarrow q)$	$\wedge$	q	$\Rightarrow$	p
T	T	T	T	T	T	T
T	F	F	F	F	T	T
F	T	T	T	T	F	F
F	F	T	F	F	T	F

The argument is invalid.

17. Let p = "I cut the grass." Let q = "It does not rain."

$p \leftrightarrow q$

$\dfrac{q}{\therefore p}$

p	q	$(p \leftrightarrow q)$	$\wedge$	q	$\Rightarrow$	p
T	T	T	T	T	T	T
T	F	F	F	F	T	T
F	T	F	F	T	T	F
F	F	T	F	F	T	F

The argument is valid.

19. Let p = "I did not study." Let q = "I passed the exam."

$p \vee q$

$\dfrac{p}{\therefore \sim q}$

p	q	$(p \vee q)$	$\wedge$	p	$\Rightarrow$	$\sim q$
T	T	T	T	T	F	F
T	F	T	T	T	T	T
F	T	T	F	F	T	F
F	F	F	F	F	T	T

The argument is invalid.

21. Let p = "You do 20 pushups." Let q = "You pass physical education." Let r = "You graduate."

$p \rightarrow q$

$\dfrac{q \rightarrow r}{\therefore p \rightarrow r}$

p	q	r	$(p \to q)$	$\wedge$	$(q \to r)$	$\Rightarrow$	$p \to r$
T	T	T	T	T	T	T	T
T	T	F	T	F	F	T	F
T	F	T	F	F	T	T	T
T	F	F	F	F	T	T	F
F	T	T	T	T	T	T	T
F	T	F	T	F	F	T	T
F	F	T	T	T	T	T	T
F	F	F	T	T	T	T	T

The argument is valid.

23. An argument consists of two parts: (a) Two or more statements, called *premises*, which are joined logically by and(s). (b) A last statement, called a *conclusion*. The argument is valid if part (b) follows logically from part (a), or else it's invalid.

25. Yes; an argument is invalid if it has at least one F in the $\to$ column. Whether or not the conclusion is actually true in real life does not matter, as far as the truth table is concerned.

27. Let p = "Parents respect their children." Let
q = "Parents listen to their children."

$$p \to q$$
$$\underline{\sim q}$$
$$\therefore \sim p$$

p	q	$(p \to q)$	$\wedge$	$\sim q$	$\Rightarrow$	$\sim p$
T	T	T	F	F	T	F
T	F	F	F	T	T	F
F	T	T	F	F	T	T
F	F	T	T	T	T	T

The argument is valid.

Exercise Set 3-5

1.

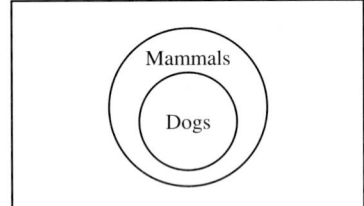

3.

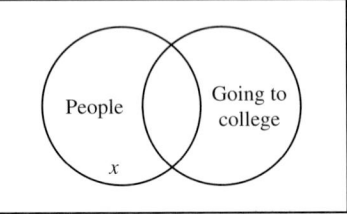

5.

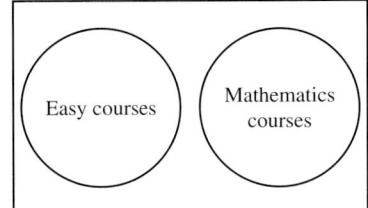

7.

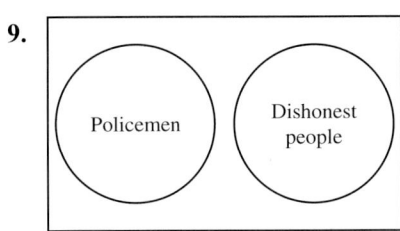

9.

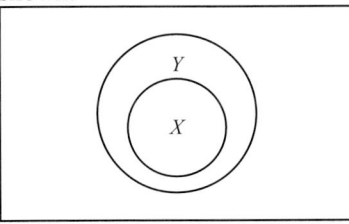

11. The first premise, "All X is Y," is diagrammed as shown.

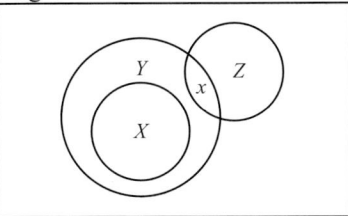

The second premise, "some Y is Z," can be diagrammed as shown.

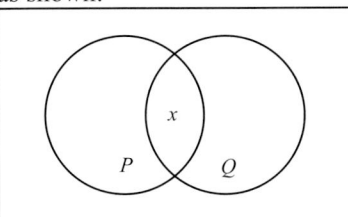

Since the diagram shows a way that the premises contradict the conclusion, "some x is z," the argument is invalid.

13. The first premise, "Some P is Q," is diagrammed as shown.

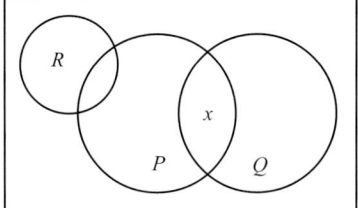

The second premise, "No Q is R," can be diagrammed as shown.

or

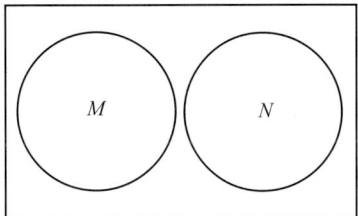

Since the element of Q that is also an element of P cannot be an element of R in either diagram, the argument is valid.

15. The first premise, "No M is N," is diagrammed as shown.

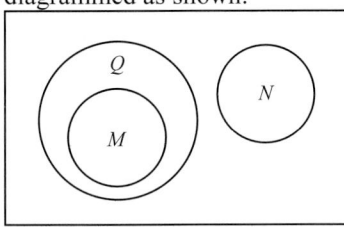

The second premise, "No N is O," can be diagrammed as shown.

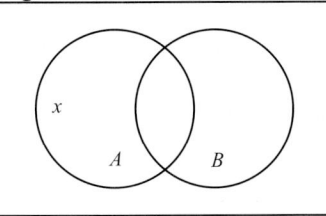

Since the diagram shows a way that the premises contradict the conclusion, "Some M is not O," the argument is invalid.

17. The first premise, "Some A is not B," is diagrammed as shown.

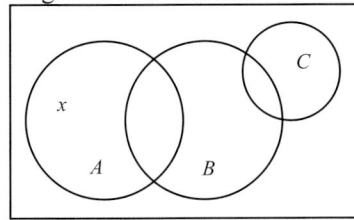

The second premise, "No A is C," can be diagrammed as shown.

or

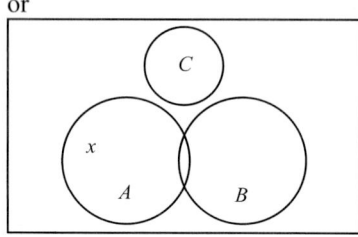

or

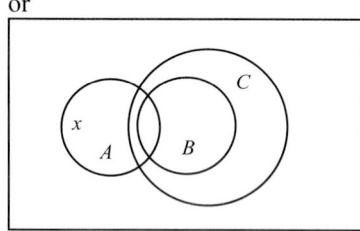

Since the element of *A* shown is not an element of *C* in any possible scenario, the conclusion, "Some *A* is not *C*," is true and the argument is valid.

19. The first premise, "No *S* is *T*" is diagrammed as shown.

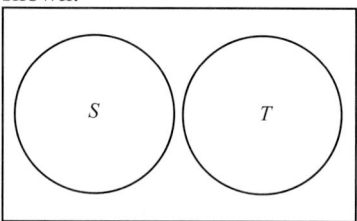

The second premise, "No *T* is *R*" can be diagrammed as shown.

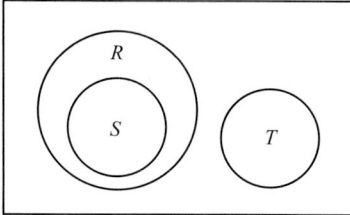

Since the diagram shows a way that the premises contradict the conclusion, "No *S* is *R*," the argument is invalid.

21. The first premise, "All fathers are men," is diagrammed as shown.

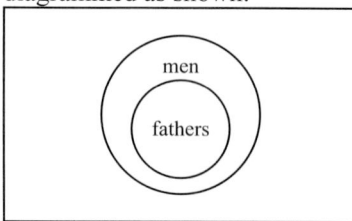

The second premise, "Some men are wealthy," can be diagrammed as shown.

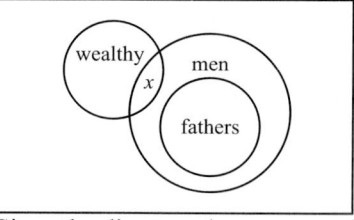

Since the diagram shows a way that the premises contradict the conclusion, "Some fathers are wealthy," the argument is invalid.

23. The first premise, "Some windstorms are violent," is diagrammed as shown.

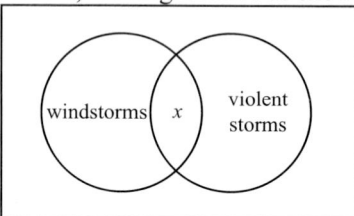

The second premise, "No snowstorms are violent," can be diagrammed as shown.

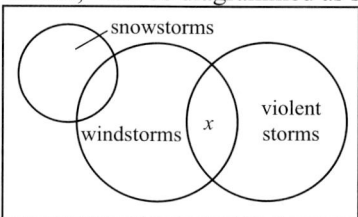

Since the diagram shows a way that the premises contradict the conclusion, "No snowstorms are windy," the argument is invalid.

25. The first premise, "Some nurses are patient," is diagrammed as shown.

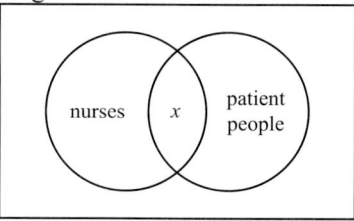

The second premise, "No patient people are rude," can be diagrammed as shown.

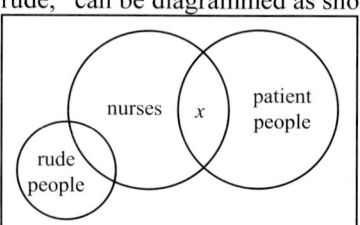

or

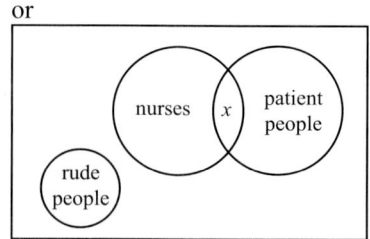

or

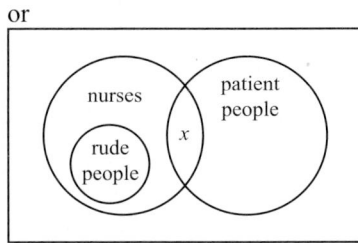

Since the element x in the nurses circle is not in the rude people circle in any possible scenario, the conclusion, "Some nurses are not rude," is true and the argument is valid.

27. The first premise, "Some gamblers are poor," is diagrammed as shown.

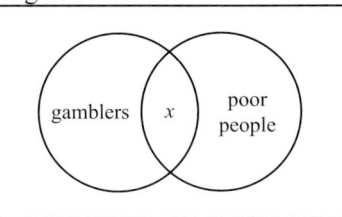

The second premise, "No gamblers are intelligent," can be diagrammed as shown.

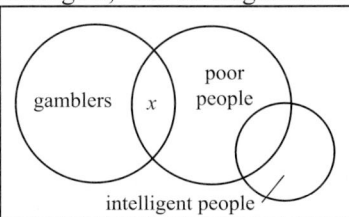

Since the diagram shows a way that the premises contradict the conclusion, "No poor people are intelligent," the argument is invalid.

29. The first premise, "Some machines have gears," is diagrammed as shown.

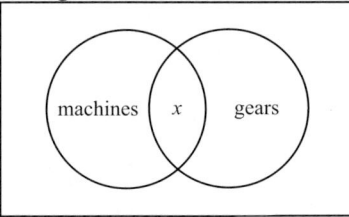

The second premise, "Some vehicles have gears," can be diagrammed as shown.

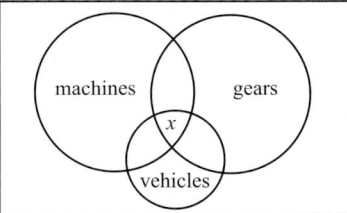

Since the diagram shows a way that the premises contradict the conclusion, "No machines are vehicles," the argument is invalid.

31. (a) Universal Affirmative: All computers have a memory.

   (b) Universal Negative: No chickens have horns.

   (c) Particular Affirmative: Some mountains always have snow.

   (d) Particular Negative: Some trees do not bear fruit.

33. All A is C.

35. No birds have teeth.

Review Exercises

1. $p \wedge q$

3. $q \leftrightarrow p$

5. $\sim p \rightarrow \sim q$

7. $\sim(p \rightarrow q)$

9. $\sim(\sim q)$

11. It is cool or not cloudy.

13. It is cool if and only if it is cloudy.

15. It's not the case that it is not cool, or it is cloudy.

17.

p	q	r	$(p$	$\rightarrow$	$\sim q)$	$\vee$	r
T	T	T	T	F	F	T	T
T	T	F	T	F	F	F	F
T	F	T	T	T	T	T	T
T	F	F	T	T	T	T	F
F	T	T	F	T	F	T	T
F	T	F	F	T	F	T	F
F	F	T	F	T	T	T	T
F	F	F	F	T	T	T	F
			①	③	②	⑤	④

19.

p	q	$\sim p$	$\rightarrow$	$(\sim q$	$\vee$	$p)$
T	T	F	T	F	T	T
T	F	F	T	T	T	T
F	T	T	F	F	F	F
F	F	T	T	T	T	F
		①	⑤	②	④	③

21.

p	q	p	→	(p ∨ q)
T	T	T	T	T
T	F	T	T	T
F	T	F	T	T
F	F	F	T	F
			↑	

Since the truth table value consists of all Ts, the statement is a tautology.

23.

p	q	(p	∧	~q)	↔	(q	∧	~p)
T	T	T	F	F	T	T	F	F
T	F	T	T	T	F	F	F	F
F	T	F	F	F	F	T	T	T
F	F	F	F	T	T	F	F	T
					↑			

Since the statement can be true in some cases and false in some cases, it is neither a tautology nor a self-contradiction.

25.

p	q	(~q	∨	p)	∧	q
T	T	F	T	T	T	T
T	F	T	T	T	F	F
F	T	F	F	F	F	T
F	F	T	T	F	F	F
					↑	

Since the statement can be true in some cases and false in some cases, it is neither a tautology nor a self-contradiction.

27.

p	q	~p	∨	~q	~	(p ↔ q)
T	T	F	F	F	F	T
T	F	F	T	T	T	F
F	T	T	T	F	T	F
F	F	T	T	T	F	T
			↑		↑	

The statements are not logically equivalent.

29.

p	q	(p	→	~q)	∧	(~q	↔	~p)	⇒	p
T	T	T	F	F	F	F	T	F	T	T
T	F	T	T	T	F	T	F	F	T	T
F	T	F	T	F	F	F	F	T	T	F
F	F	F	T	T	T	T	T	T	F	F

The argument is invalid.

31.

p	q	r	(~p	∨	q)	∧	(q	∨	~r)	⇒	q	→	(~p	∧	~r)
T	T	T	F	T	T	T	T	T	F	F	T	F	F	F	F
T	T	F	F	T	T	T	T	T	T	F	T	F	F	F	T
T	F	T	F	F	F	F	F	F	F	T	F	T	F	F	F
T	F	F	F	F	F	F	F	T	T	T	F	T	F	F	T
F	T	T	T	T	T	T	T	T	F	F	T	F	T	F	F
F	T	F	T	T	T	T	T	T	T	T	T	T	T	T	T
F	F	T	T	T	F	F	F	F	F	T	F	T	T	F	F
F	F	F	T	T	F	T	F	T	T	T	F	T	T	T	T

The argument is invalid.

33. *Converse*: If it says "Meow," then it is a cat.
Inverse: If it is not a cat, then it does not say, "Meow."
Contrapositive: If it does not say, "Meow," then it is not a cat.

35. *Converse*: If the car is not very fast, then it is red.
Inverse: If the car is not red, then it is very fast.
Contrapositive: If the car is very fast, then it is not red.

37. The first premise, "Some *A* is not *C*," is diagrammed as shown.

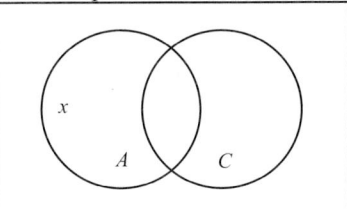

The second premise, "Some *B* is not *C*," can be diagrammed as shown.

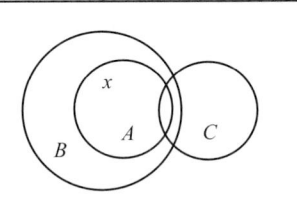

Since the diagram shows a way that the premises contradict the conclusion, "Some *A* is not *B*," the argument is invalid.

Chapter Test

1. $p \wedge q$

3. $p \leftrightarrow q$

5. $\sim(\sim p \wedge q)$

7. It is sunny or not warm.

9. It is sunny if and only if it is warm.

11. It is not the case that it is not sunny or it is warm.

13.

p	q	r	(p	→	~q)	∧	r
T	T	T	T	F	F	F	T
T	T	F	T	F	F	F	F
T	F	T	T	T	T	T	T
T	F	F	T	T	T	F	F
F	T	T	F	T	F	T	T
F	T	F	F	T	F	F	F
F	F	T	F	T	T	T	T
F	F	F	F	T	T	F	F
			①	③	②	⑤	④

15.

p	q	(~q	∨	p)	∧	p
T	T	F	T	T	T	T
T	F	T	T	T	T	T
F	T	F	F	F	F	F
F	F	T	T	F	F	F
		①	③	②	⑤	④

17.

p	q	(p ∨ q)	∧	p
T	T	T	T	T
T	F	T	T	T
F	T	T	F	F
F	F	F	F	F
			↑	

Since the statement can be true in some cases and false in some cases, it is neither a tautology nor a self-contradiction.

19.

p	q	(p	∨	~q)	↔	(p	→	~q)
T	T	T	T	F	F	T	F	F
T	F	T	T	T	T	T	T	T
F	T	F	F	F	F	F	T	F
F	F	F	T	T	T	F	T	T
					↑			

Since the statement can be true in some cases and false in some cases, it is neither a tautology nor a self-contradiction.

21.

p	q	$(p \wedge q)$	$\wedge$	p
T	T	T	T	T
T	F	F	F	T
F	T	F	F	F
F	F	F	F	F

Since the statement can be true in some cases and false in some cases, it is neither a tautology nor a self contradiction.

23.

p	q	r	$(p$	$\rightarrow$	$q)$	$\wedge$	$(\sim q$	$\vee$	$\sim r)$	$\Rightarrow$	q	$\leftrightarrow$	$(\sim p$	$\wedge$	$\sim r)$
T	T	T	T	T	T	F	F	F	F	T	T	F	F	F	F
T	T	F	T	T	T	T	F	T	T	F	T	F	F	F	T
T	F	T	T	F	F	F	T	T	F	T	F	T	F	F	F
T	F	F	T	F	F	F	T	T	T	T	F	T	F	F	T
F	T	T	F	T	T	F	F	F	F	T	T	F	T	F	F
F	T	F	F	T	T	T	F	T	T	T	T	T	T	T	T
F	F	T	F	T	F	T	T	T	F	T	F	T	T	F	F
F	F	F	F	T	F	T	T	T	T	F	F	F	T	T	T

The argument is invalid.

25.

p	q	r	$(\sim p$	$\rightarrow$	$\sim r)$	$\wedge$	$(\sim r$	$\vee$	$\sim q)$	$\Rightarrow$	q	$\leftrightarrow$	p
T	T	T	F	T	F	F	F	F	F	T	T	T	T
T	T	F	F	T	T	T	T	T	F	T	T	T	T
T	F	T	F	T	F	T	F	T	T	F	F	F	T
T	F	F	F	T	T	T	T	T	T	F	F	F	T
F	T	T	T	F	F	F	F	F	F	T	T	F	F
F	T	F	T	T	T	T	T	T	F	F	T	F	F
F	F	T	T	F	F	F	F	T	T	T	F	T	F
F	F	F	T	T	T	T	T	T	T	T	F	T	F

The argument is invalid.

27.

p	q	r	$(p \vee q)$	$\wedge$	r	$(p \wedge r)$	$\vee$	$(q \wedge r)$
T	T	T	T	T	T	T	T	T
T	T	F	T	F	F	F	F	F
T	F	T	T	T	T	T	T	F
T	F	F	T	F	F	F	F	F
F	T	T	T	T	T	F	T	T
F	T	F	T	F	F	F	F	F
F	F	T	F	F	T	F	F	F
F	F	F	F	F	F	F	F	F
				↑			↑	

The statements are logically equivalent.

29. The first premise, "No *B* is *A*," is diagrammed as shown.

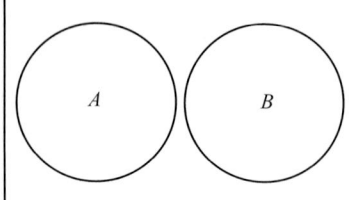

The second premise, "Some *A* is *C*," can be diagrammed as shown.

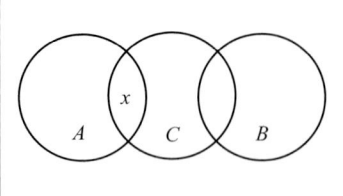

Since the diagram shows a way that the premises contradict the conclusion, "No *B* is *C*," the argument is invalid.

4 | Numeration Systems

Exercise Set 4-1

1. The number consists of 3 tens and 5 ones; hence it equals $3 \times 10 + 5 \times 1 = 35$.

3. The number consists of 2 ten thousands, 2 hundreds, 2 tens, and 5 ones; hence it equals
$2 \times 10{,}000 + 2 \times 100 + 2 \times 10 + 5 \times 1 = 20{,}225$.

5. The number consists of 3 ten thousands,
1 hundred, 6 tens, and 3 ones; hence it equals 30,163.

7. The number consists of 2 ten thousands,
3 hundreds, 1 ten and 4 ones; hence it equals
$2 \times 10{,}000 + 3 \times 100 + 1 \times 10 + 4 \times 1 = 20{,}314$.

9. The number consists of 1 million, 1 hundred thousand, 1 ten thousand, 2 thousands, and
1 ten; hence it equals
$1 \times 1{,}000{,}000 + 1 \times 100{,}000 + 1 \times 10{,}000$
$+ 2 \times 1000 + 1 \times 10 = 1{,}112{,}010$.

11. Since 7 consists of 7 ones, it is written as │││││││.

13. Since 37 consists of 3 tens and 7 ones, it is written as ∩∩∩│││││││.

15. Since 168 consists of 1 hundred, 6 tens, and
8 ones, it is written as ୨∩∩∩∩∩∩│││││││.

17. Since 801 consists of 8 hundreds and 1 one, it is written as ୨୨୨୨୨୨୨୨│.

19. Since 1256 consists of 1 thousand, 2 hundreds,
5 tens, and 6 ones, it is written as ꜙ୨୨∩∩∩∩∩││││││.

21. There are 11 heel bones and 6 vertical staffs. Replace 10 heel bones with a scroll. The final answer is
୨∩││││││.

23. There are 3 fish, 5 pointing fingers, 3 heel bones, and 5 vertical staffs. The final answer is
◁◁◁ⲅⲅⲅⲅⲅ∩∩∩│││││.

25. Convert 1 scroll to 9 heel bones and 10 vertical staffs. Then convert 1 pointing finger to 9 lotus flowers and 10 scrolls. The final answer is shown.

  ⲅꜙꜙꜙꜙꜙꜙꜙꜙꜙ୨୨୨୨୨୨୨୨୨
  ∩∩∩∩∩∩∩∩∩│││││││││││││││
  − ⲅ୨୨୨││││
  _____
  ꜙꜙꜙꜙꜙꜙꜙꜙꜙ୨୨୨୨୨୨୨୨
  ∩∩∩∩∩∩∩∩∩│││││││││││

27. Since there are 1 ten and 2 ones, the number represents 12.

29. Since there are 5 tens and 1 one, the number represents 51.

31. There are 11 3600s, 21 60s, and 11 1s. The number is

$$11 \times 3600 = 39,600$$
$$21 \times 60 = 1,260$$
$$\underline{11 \times 1 = 11}$$
$$40,871$$

33. There are 30 3600s, 21 60s, and 24 12. The number is

$$30 \times 3600 = 108,000$$
$$21 \times 60 = 1,260$$
$$\underline{24 \times 1 = 24}$$
$$109,284$$

35. There are 13 60s and 12 1s. The number is

$$13 \times 60 = 780$$
$$\underline{12 \times 1 = 12}$$
$$792$$

37. ⟨⟨⟨ ▼▼

39. $78 \div 60 = 1$ remainder 18

▼   ⟨▼▼▼▼▼▼▼▼

41. $292 \div 60 = 4$ remainder 52

▼▼▼▼   ⟨⟨⟨⟨⟨▼▼

43. $1023 \div 60 = 17$ remainder 3

⟨▼▼▼▼▼▼▼   ▼▼▼

45. $5216 \div 3600 = 1$ remainder 1616
$1616 \div 60 = 26$ remainder 56

▼   ⟨⟨▼▼▼▼▼▼   ⟨⟨⟨⟨⟨▼▼▼▼▼▼

47. X = 10, V = 5, and II = 2; hence XVII = 17.

49. XL = 40 and III = 3; hence XLIII = 43.

51. L = 50, XXX = 30, V = 5, and I = 1; hence LXXXVI = 86.

53. CD = 400, X = 10, V = 5, and III = 3; hence CDXVII = 418.

55. CD = 400 and XC = 90; hence CDXC = 490.

57. 39 is written as 30 + 9 or XXXIX.

59. 567 is written as 500 + 50 + 10 + 7 or DLXVII.

61. 1258 is written as 1000 + 200 + 50 + 8 or MCCLVIII.

63. 1462 is written as 1000 + 400 + 50 + 10 + 2 or MCDLXII.

65. 3000 is written as MMM.

67. hundreds

69. ten thousands

71. ones

73. $86 = 80 + 6$
$ = 8 \times 10^1 + 6$

75. $1812 = 1000 + 800 + 10 + 2$
$= 1 \times 1000 + 8 \times 100 + 1 \times 10 + 2$
$= 1 \times 10^3 + 8 \times 10^2 + 1 \times 10^1 + 2$

77. $6002 = 6 \times 1000 + 2$
$= 6 \times 10^3 + 2$

79. $162,873 = 100,000 + 60,000 + 2,000 + 800 + 70 + 3$
$= 1 \times 100,000 + 6 \times 10,000 + 2 \times 1,000 + 8 \times 100 + 7 \times 10 + 3$
$= 1 \times 10^5 + 6 \times 10^4 + 2 \times 10^3 + 8 \times 10^2 + 7 \times 10^1 + 3$

81. $17,531,801 = 10,000,000 + 7,000,000 + 500,000 + 30,000 + 1,000 + 800 + 1$
$= 1 \times 10,000,000 + 7 \times 1,000,000 + 5 \times 100,000 + 3 \times 10,000 + 1 \times 1,000 + 8 \times 100 + 1$
$= 1 \times 10^7 + 7 \times 10^6 + 5 \times 10^5 + 3 \times 10^4 + 1 \times 10^3 + 8 \times 10^2 + 1$

83. M = 1000, CM = 900, XXX = 30, and IX = 9; hence MCMXXXIX = 1939.

85. M = 1000, CM = 900, XC = 90, V = 5 and II = 2; hence MCMXCVII = 1997.

87. Each symbol has a certain value (= number); we add these values to obtain the total value.

89. It's primarily an additive system, in the sense that we add values to obtain the total; however, the letters must be in the correct relative position. Example: CCX = 100 + 100 + 10 = 210, but CXC = 100 + 90 = 190.

91. Answers will vary.

93. Answers will vary.

Exercise Set 4-2

1. $1011_{two} = 1 \times 2^3 + 0 \times 2^2 + 1 \times 2^1 + 1$
$= 1 \times 8 + 0 \times 4 + 1 \times 2 + 1$
$= 8 + 0 + 2 + 1 = 11$

3. $53_{six} = 5 \times 6^1 + 3$
$= 5 \times 6 + 3$
$= 30 + 3 = 33$

5. $99_{eleven} = 9 \times 11^1 + 9$
$= 9 \times 11 + 9$
$= 99 + 9 = 108$

7. $10221_{three} = 1 \times 3^4 + 0 \times 3^3 + 2 \times 3^2 + 2 \times 3^1 + 1$
$= 1 \times 81 + 0 \times 27 + 2 \times 9 + 2 \times 3 + 1$
$= 81 + 0 + 18 + 6 + 1 = 106$

9. $2221_{five} = 2 \times 5^3 + 2 \times 5^2 + 2 \times 5^1 + 1$
$= 2 \times 125 + 2 \times 25 + 2 \times 5 + 1$
$= 250 + 50 + 10 + 1 = 311$

11. $153_{six} = 1 \times 6^2 + 5 \times 6^1 + 3$
$= 1 \times 36 + 5 \times 6 + 3$
$= 36 + 30 + 3 = 69$

13. $438_{nine} = 4 \times 9^2 + 3 \times 9^1 + 8$
$\qquad = 4 \times 81 + 3 \times 9 + 8$
$\qquad = 324 + 27 + 8 = 359$

15. $352_{seven} = 3 \times 7^2 + 5 \times 7^1 + 2$
$\qquad = 3 \times 49 + 5 \times 7 + 2$
$\qquad = 147 + 35 + 2 = 184$

17. $921E_{sixteen} = 9 \times 16^3 + 2 \times 16^2 + 1 \times 16^1 + 14$
$\qquad = 9 \times 4096 + 2 \times 256 + 1 \times 16 + 14$
$\qquad = 36,864 + 512 + 16 + 14 = 37,406$

19. $812_{twelve} = 8 \times 12^2 + 1 \times 12^1 + 2$
$\qquad = 8 \times 144 + 1 \times 12 + 2$
$\qquad = 1152 + 12 + 2 = 1166$

21. The place values for base two are 1, 2, 4, 8, 16, etc.

$$16\overline{)31} \quad 8\overline{)15} \quad 4\overline{)7} \quad 2\overline{)3}$$
with quotients 1, 1, 1, 1 and remainders:
$$\frac{16}{15} \quad \frac{8}{7} \quad \frac{4}{3} \quad \frac{2}{1}$$

$31 = 11111_{two}$

23. The place values for base six are 1, 6, 36, 216, 1296, etc.

$$216\overline{)345} \quad 36\overline{)129} \quad 6\overline{)21}$$
quotients 1, 3, 3:
$$\frac{216}{129} \quad \frac{108}{21} \quad \frac{18}{3}$$

$345 = 1333_{six}$

25. The place values for base seven are 1, 7, 49, 343, etc.

$$7\overline{)16}$$
quotient 2:
$$\frac{14}{2}$$

$16 = 22_{seven}$

27. The place values for base nine are 1, 9, 81, 729, 6561, etc.

$$729\overline{)745} \quad 81\overline{)16} \quad 9\overline{)16}$$
quotients 1, 0, 1:
$$\frac{729}{16} \quad \frac{0}{16} \quad \frac{9}{7}$$

$745 = 1017_{nine}$

29. The place values for base two are 1, 2, 4, 8, 16, 32, etc.

$$16\overline{)22} \quad 8\overline{)6} \quad 4\overline{)6} \quad 2\overline{)2}$$
quotients 1, 0, 1, 1:
$$\frac{16}{6} \quad \frac{0}{6} \quad \frac{4}{2} \quad \frac{2}{0}$$

$22 = 10110_{two}$

31. The place values for base five are 1, 5, 25, 125, etc.

$$5\overline{)18}$$
quotient 3:
$$\frac{15}{3}$$

$18 = 33_{five}$

33. The place values for base sixteen are 1, 16, 256, 4096, etc.

$$256\overline{)2361} \quad 16\overline{)57}$$
quotients 9, 3:
$$\frac{2304}{57} \quad \frac{48}{9}$$

$2361 = 939_{sixteen}$

35. The place values for base five are 1, 5, 25, 125, 625, 3125, 15625, etc.

$$15625\overline{)18432} \quad 3125\overline{)2807} \quad 625\overline{)2807}$$
quotients 1, 0, 4:
$$\frac{15625}{2807} \quad \frac{0}{2807} \quad \frac{2500}{307}$$

$$125\overline{)307} \quad 25\overline{)57} \quad 5\overline{)7}$$
quotients 2, 2, 1:
$$\frac{250}{57} \quad \frac{50}{7} \quad \frac{5}{2}$$

$18,432 = 1042212_{five}$

37. The place values for base seven are 1, 7, 49, etc.

$$7\overline{)32}$$
quotient 4:
$$\frac{28}{4}$$

$32 = 44_{seven}$

39. The place values for base two are 1, 2, 4, 8, 16, 32, 64, 128, 256, etc.

$$256\overline{)256} \quad 128\overline{)0} \quad 64\overline{)0} \quad 32\overline{)0} \quad 16\overline{)0}$$
quotients 1, 0, 0, 0, 0:
$$\frac{256}{0} \quad \frac{0}{0} \quad \frac{0}{0} \quad \frac{0}{0} \quad \frac{0}{0}$$

$$8\overline{)0} \quad 4\overline{)0} \quad 2\overline{)0}$$
quotients 0, 0, 0:
$$\frac{0}{0} \quad \frac{0}{0} \quad \frac{0}{0}$$

$256 = 100000000_{two}$

41. $134_{six} = 1 \times 6^2 + 3 \times 6^1 + 4$
$\qquad = 1 \times 36 + 3 \times 6 + 4$
$\qquad = 36 + 18 + 4 = 58_{ten}$

The place values for base two are 1, 2, 4, 8, 16, 32, 64, etc.

$$\begin{array}{r}1\\32\overline{)58}\\32\\\overline{26}\end{array}\quad\begin{array}{r}1\\16\overline{)26}\\16\\\overline{10}\end{array}\quad\begin{array}{r}1\\8\overline{)10}\\8\\\overline{2}\end{array}\quad\begin{array}{r}0\\4\overline{)2}\\0\\\overline{2}\end{array}\quad\begin{array}{r}1\\2\overline{)2}\\2\\\overline{0}\end{array}$$

$58 = 111010_{\text{two}}$

43. $342_{\text{five}} = 3\times 5^2 + 4\times 5^1 + 2$
$= 3\times 25 + 4\times 5 + 2$
$= 75 + 20 + 2 = 97$

The place values for base twelve are 1, 12, 144, etc.

$$\begin{array}{r}8\\12\overline{)97}\\96\\\overline{1}\end{array}$$

$97 = 81_{\text{twelve}}$

45. $1221_{\text{three}} = 1\times 3^3 + 2\times 3^2 + 2\times 3^1 + 1$
$= 1\times 27 + 2\times 9 + 2\times 3 + 1$
$= 27 + 18 + 6 + 1 = 52_{\text{ten}}$

The place values for base four are 1, 4, 16, 64, etc.

$$\begin{array}{r}3\\16\overline{)52}\\48\\\overline{4}\end{array}\quad\begin{array}{r}1\\4\overline{)4}\\4\\\overline{0}\end{array}$$

$52 = 310_{\text{four}}$

47. $432_{\text{eight}} = 4\times 8^2 + 3\times 8^1 + 2$
$= 4\times 64 + 3\times 8 + 2$
$= 256 + 24 + 2 = 282_{\text{ten}}$

The place values for base twelve are 1, 12, 144, 1728, etc.

$$\begin{array}{r}1\\144\overline{)282}\\144\\\overline{138}\end{array}\quad\begin{array}{r}11\\12\overline{)138}\\12\\\overline{18}\\12\\\overline{6}\end{array}$$

$282 = 1B6_{\text{twelve}}$

49. $1782_{\text{nine}} = 1\times 9^3 + 7\times 9^2 + 8\times 9^1 + 2$
$= 1\times 729 + 7\times 81 + 8\times 9 + 2$
$= 729 + 567 + 72 + 2 = 1370_{\text{ten}}$

The place values for base seven are 1, 7, 49, 343, 2401, etc.

$$\begin{array}{r}3\\343\overline{)1370}\\1029\\\overline{341}\end{array}\quad\begin{array}{r}6\\49\overline{)341}\\294\\\overline{47}\end{array}\quad\begin{array}{r}6\\7\overline{)47}\\42\\\overline{5}\end{array}$$

$1370 = 3665_{\text{seven}}$

51. $$\begin{array}{r}5\\16\overline{)87}\\80\\\overline{7}\end{array}$$

87 oz = 5 lb 7 oz

53. $$\begin{array}{r}34\\36\overline{)1256}\\108\\\overline{176}\\144\\\overline{32}\end{array}\quad\begin{array}{r}2\\12\overline{)32}\\24\\\overline{8}\end{array}$$

1256 in. = 34 yd 2 ft 8 in.

55. STOP

57. CLASSISOVER

59. ITISRAINING

61. 13256

63. 67138

65. 44501

67. ꟷꟷ (barcode)

69. ꟷꟷ (barcode)

71. ꟷꟷ (barcode)

73. Let b = the base = some integer (exceeding 1). In the number to base b, the rightmost digit has a place value of $b^0 = 1$; the next digit has a place value of $b^1 = b$; the next digit has a place value of $b^2 = b\times b$; etc.

75. Let y = the given number in base ten, and let b = the "other base." Divide y by the biggest power of b (say b^m) that goes into y; let q_m be the quotient, exclusive of the remainder r_m. Divide r_m, regardless of its value, by b^{m-1}; let q_{m-1} be the quotient, exclusive of the remainder r_{m-1}. Continue the process until you have divided by b^1. The desired answer $= q_m q_{m-1} q_{m-2}\cdots q_1 r_1$.

77. 6 manufacturer's

79. 0 product

Exercise Set 4-3

1.
$$\begin{array}{r}11_{\text{five}}\\+\ 21_{\text{five}}\\\hline 32_{\text{five}}\end{array}$$

3.
$$\begin{array}{r}11\\3230_{\text{four}}\\+\ 1322_{\text{four}}\\\hline 11212_{\text{four}}\end{array}$$

$3_{\text{four}} + 2_{\text{four}} = 11_{\text{four}}$
$1_{\text{four}} + 2_{\text{four}} + 3_{\text{four}} = 12_{\text{four}}$
$1_{\text{four}} + 3_{\text{four}} + 1_{\text{four}} = 11_{\text{four}}$

5.
$$\begin{array}{r} {\scriptstyle 1\,1} \\ {\scriptstyle 1}\,321_{six} \\ +\,1255_{six} \\ \hline 2020_{six} \end{array}$$

$1_{six} + 5_{six} = 10_{six}$
$1_{six} + 2_{six} + 5_{six} = 12_{six}$
$1_{six} + 3_{six} + 2_{six} = 10_{six}$

7.
$$\begin{array}{r} 4344_{nine} \\ +\,2313_{nine} \\ \hline 6657_{nine} \end{array}$$

9.
$$\begin{array}{r} 43_{five} \\ -\,12_{five} \\ \hline 31_{five} \end{array}$$

11.
$$\begin{array}{r} 262_{seven} \\ -\,161_{seven} \\ \hline 101_{seven} \end{array}$$

13.
$$\begin{array}{r} {\scriptstyle 7\ 13} \\ 42\cancel{8}\cancel{1}_{nine} \\ -\,2781_{nine} \\ \hline 40040_{nine} \end{array}$$

$13_{nine} - 8_{nine} = 4_{nine}$

15.
$$\begin{array}{r} 12AB_{twelve} \\ -\,93A_{twelve} \\ \hline 571_{twelve} \end{array}$$

$B_{twelve} - A_{twelve} = 1_{twelve}$
$A_{twelve} - 3_{twelve} = 7_{twelve}$
$12_{twelve} - 9_{twelve} = 5_{twelve}$

17.
$$\begin{array}{r} {\scriptstyle 1} \\ 52_{six} \\ \times\,4_{six} \\ \hline 332_{six} \end{array}$$

$2_{six} \times 4_{six} = 12_{six}$
$5_{six} \times 4_{six} + 1_{six} = 33_{six}$

19.
$$\begin{array}{r} 818_{nine} \\ \times\,62_{nine} \\ \hline 1737_{nine} \\ 54230_{nine} \\ \hline 56067_{nine} \end{array}$$

$8_{nine} \times 2_{nine} = 17_{nine}$
$8_{nine} \times 6_{nine} = 53_{nine}$
$1_{nine} \times 6_{nine} + 5_{nine} = 12_{nine}$
$8_{nine} \times 6_{nine} + 1_{nine} = 54_{nine}$

21.
$$\begin{array}{r} AB5_{twelve} \\ \times\,42_{twelve} \\ \hline 19AA_{twelve} \\ 3798\,0_{twelve} \\ \hline 3976A_{twelve} \end{array}$$

$5_{twelve} \times 2_{twelve} = A_{twelve}$
$B_{twelve} \times 2_{twelve} = 1A_{twelve}$
$A_{twelve} \times 2_{twelve} + 1_{twelve} = 19_{twelve}$
$5_{twelve} \times 4_{twelve} = 18_{twelve}$
$B_{twelve} \times 4_{twelve} + 1_{twelve} = 39_{twelve}$
$A_{twelve} \times 4_{twelve} + 3_{twelve} = 37_{twelve}$
$A_{twelve} + 8_{twelve} = 16_{twelve}$
$1_{twelve} + 9_{twelve} + 9_{twelve} = 17_{twelve}$

23.
$$\begin{array}{r} 482_{nine} \quad \text{remainder } 2_{nine} \\ 3_{nine}\overline{)1568_{nine}} \\ \underline{13} \\ 26 \\ \underline{26} \\ 08 \\ \underline{6} \\ 2 \end{array}$$

25.
$$\begin{array}{r} 230_{five} \quad \text{remainder } 3_{five} \\ 4_{five}\overline{)2023_{five}} \\ \underline{13} \\ 22 \\ \underline{22} \\ 03 \\ \underline{0} \\ 3 \end{array}$$

27.
$$\begin{array}{r} {\scriptstyle 1\,1\,1} \\ 1001_{two} \\ +\,111_{two} \\ \hline 10000_{two} \end{array}$$

29.
$$\begin{array}{r} {\scriptstyle 1\,1} \\ 3BA_{sixteen} \\ +\,49_{sixteen} \\ \hline 403_{sixteen} \end{array}$$

31.
$$\begin{array}{r} {\scriptstyle 0\ 1\,1\ 10} \\ 1\cancel{1}\,\cancel{0}\,\cancel{0}_{two} \\ -\,11_{two} \\ \hline 1001_{two} \end{array}$$

33.
$$\begin{array}{r} {\scriptstyle 4\ 11\,16} \\ \cancel{5}\,\cancel{2}\,\cancel{6}B_{sixteen} \\ -\,4\ A1_{sixteen} \\ \hline 4DCA_{sixteen} \end{array}$$

35.
$$\begin{array}{r} 1010_{two} \\ \times\,101_{two} \\ \hline 1010_{two} \\ 101000_{two} \\ \hline 110010_{two} \end{array}$$

37.
$$\begin{array}{r} {\scriptstyle 1} \\ A25_{sixteen} \\ \times\,4_{sixteen} \\ \hline 2894_{sixteen} \end{array}$$

39.
$$\begin{array}{r} 11_{two} \quad \text{remainder } 10_{two} \\ 11_{two}\overline{)1011_{two}} \\ \underline{11} \\ 101 \\ \underline{11} \\ 10 \end{array}$$

41.

$$\begin{array}{r} B23_{\text{sixteen}} \quad \text{remainder } 2_{\text{sixteen}} \\ 5_{\text{sixteen}}\overline{)37B1_{\text{sixteen}}} \\ \underline{37} \\ 0B \\ \underline{A} \\ 11 \\ \underline{F} \\ 2 \end{array}$$

43. Consider $a - b$ in base eight. Find b at extreme left column of the table, and go horizontally to a. From here, go up to the top row of the table; the entry you meet is the answer. The idea is to "go backward" in the addition table because subtraction is the opposite of addition.

45. L

47. G

49. 0101 0011 0101 0100 0100 1111 0101 0000

51. 0100 1000 0100 0101 0100 1100 0100 1100 0100 1111

Review Exercises

1. The number consists of 1 million, 2 hundreds, 2 tens, and 1 one; hence it equals 1,000,221.

3. There are 11 60s and 21 1s. The number is
$$\begin{array}{rcl} 11 \times 60 &=& 660 \\ 21 \times 1 &=& 21 \\ \hline && 681 \end{array}$$

5. CD = 400, X = 10, IX = 9; hence CDXIX = 419.

7. 896 is written as 500 + 300 + 90 + 6 or DCCCXCVI.

9. Since 125 consists of 1 hundred, 2 tens, and 5 ones, it is written as ⑨∩∩|||||.

11. 1110111_{two}
$$= 1 \times 2^6 + 1 \times 2^5 + 1 \times 2^4 + 0 \times 2^3 + 1 \times 2^2 + 1 \times 2^1 + 1$$
$$= 1 \times 64 + 1 \times 32 + 1 \times 16 + 0 \times 8 + 1 \times 4 + 1 \times 2 + 1$$
$$= 64 + 32 + 16 + 0 + 4 + 2 + 1 = 119$$

13. $A03B_{\text{twelve}} = 10 \times 12^3 + 0 \times 12^2 + 3 \times 12^1 + 11$
$$= 10 \times 1728 + 0 \times 144 + 3 \times 12 + 11$$
$$= 17,280 + 0 + 36 + 11 = 17,327$$

15. $14441_{\text{five}} = 1 \times 5^4 + 4 \times 5^3 + 4 \times 5^2 + 4 \times 5^1 + 1$
$$= 1 \times 625 + 4 \times 125 + 4 \times 25 + 4 \times 5 + 1$$
$$= 625 + 500 + 100 + 20 + 1 = 1246$$

17. $6000_{\text{seven}} = 6 \times 7^3 + 0 \times 7^2 + 0 \times 7^1 + 0$
$$= 6 \times 343 + 0 \times 49 + 0 \times 7 + 0$$
$$= 2058 + 0 + 0 + 0 = 2058$$

19. $555_{\text{six}} = 5 \times 6^2 + 5 \times 6^1 + 5$
$$= 5 \times 36 + 5 \times 6 + 5$$
$$= 180 + 30 + 5 = 215$$

21. The place values for base six are 1, 6, 36, etc.
$$\begin{array}{r} 5 \\ 6\overline{)32} \\ \underline{30} \\ 2 \end{array}$$
$32 = 52_{\text{six}}$

23. The place values for base nine are 1, 9, 81, 729, 6561, etc.
$$\begin{array}{r} 2 \\ 729\overline{)2001} \\ \underline{1458} \\ 543 \end{array} \quad \begin{array}{r} 6 \\ 81\overline{)543} \\ \underline{486} \\ 57 \end{array} \quad \begin{array}{r} 6 \\ 9\overline{)57} \\ \underline{54} \\ 3 \end{array}$$
$2001 = 2663_{\text{nine}}$

25. The place values for base two are 1, 2, 4, 8, 16, 32, 64, etc.
$$\begin{array}{r} 1 \\ 32\overline{)43} \\ \underline{32} \\ 11 \end{array} \quad \begin{array}{r} 0 \\ 16\overline{)11} \\ \underline{0} \\ 11 \end{array} \quad \begin{array}{r} 1 \\ 8\overline{)11} \\ \underline{8} \\ 3 \end{array} \quad \begin{array}{r} 0 \\ 4\overline{)3} \\ \underline{0} \\ 3 \end{array} \quad \begin{array}{r} 1 \\ 2\overline{)3} \\ \underline{2} \\ 1 \end{array}$$
$43 = 101011_{\text{two}}$

27. The place values for base four are 1, 4, 16, 64, etc.
$$\begin{array}{r} 1 \\ 16\overline{)19} \\ \underline{16} \\ 3 \end{array} \quad \begin{array}{r} 0 \\ 4\overline{)3} \\ \underline{0} \\ 3 \end{array}$$
$19 = 103_{\text{four}}$

29. The place values for base seven are 1, 7, 49, 343, etc.
$$\begin{array}{r} 1 \\ 343\overline{)343} \\ \underline{343} \\ 0 \end{array} \quad \begin{array}{r} 0 \\ 49\overline{)0} \\ \underline{0} \\ 0 \end{array} \quad \begin{array}{r} 0 \\ 7\overline{)0} \\ \underline{0} \\ 0 \end{array}$$
$343 = 1000_{\text{seven}}$

31.
$$\begin{array}{r} 1\,1 \\ 156_{\text{nine}} \\ + \ 84_{\text{nine}} \\ \hline 251_{\text{nine}} \end{array}$$

33.
$$\begin{array}{r} 1 \\ 101110_{\text{two}} \\ + \ 1101_{\text{two}} \\ \hline 111011_{\text{two}} \end{array}$$

35.
$$\begin{array}{r} 6A20_{\text{twelve}} \\ + B096_{\text{twelve}} \\ \hline 15AB6_{\text{twelve}} \end{array}$$

37.
$$
\begin{array}{r}
{\scriptstyle 0\;1\;10} \\
10\cancel{1}\cancel{0}\;\cancel{0}11_{\text{two}} \\
-\;100\;\;111_{\text{two}} \\
\hline
101100_{\text{two}}
\end{array}
$$

39.
$$
\begin{array}{r}
{\scriptstyle 2\;12\;11} \\
\cancel{3}\;\cancel{3}\;\;\cancel{1}2_{\text{four}} \\
-\;2\;3\;\;\;21_{\text{four}} \\
\hline
3\;\;\;3\;1_{\text{four}}
\end{array}
$$

41.
$$
\begin{array}{r}
371_{\text{nine}} \\
\times\;\;\;51_{\text{nine}} \\
\hline
371_{\text{nine}} \\
20850_{\text{nine}} \\
\hline
21331_{\text{nine}}
\end{array}
$$

43.
$$
\begin{array}{r}
1101_{\text{two}} \\
\times\;\;\;111_{\text{two}} \\
\hline
1101_{\text{two}} \\
11010_{\text{two}} \\
110100_{\text{two}} \\
\hline
1011011_{\text{two}}
\end{array}
$$

45.
$$
\begin{array}{r}
230_{\text{five}} \quad \text{remainder } 2_{\text{five}} \\
3_{\text{five}}\overline{)1242_{\text{five}}} \\
\underline{11} \\
14 \\
\underline{14} \\
02 \\
\underline{0} \\
2
\end{array}
$$

47.
$$
\begin{array}{r}
342_{\text{eight}} \quad \text{remainder } 6_{\text{eight}} \\
10_{\text{eight}}\overline{)3426_{\text{eight}}} \\
\underline{30} \\
42 \\
\underline{40} \\
26 \\
\underline{20} \\
6
\end{array}
$$

Chapter Test

1. The number consists of 2 millions, 3 hundreds, 1 ten, and 2 ones; hence it equals 2,000,312.

3. There are 21 60s and 11 1s. The number is
$$
\begin{array}{rcr}
21 \times 60 &=& 1260 \\
11 \times 1 &=& 11 \\
\hline
&& 1271
\end{array}
$$

5. CD = 400, XX = 20, V = 5, and I = 1; hence CDXXVI = 426.

7. 567 is written as 500 + 50 + 10 + 7 or DLXVII.

9. Since 521 consists of 5 hundreds, 2 tens, and 1 one, it is written as ⟨⟨⟨⟨⟨∩∩I.

11. $341_{\text{five}} = 3 \times 5^2 + 4 \times 5^1 + 1$
$= 3 \times 25 + 4 \times 5 + 1$
$= 75 + 20 + 1 = 96$

13. $\text{A07B}_{\text{twelve}} = 10 \times 12^3 + 0 \times 12^2 + 7 \times 12^1 + 11$
$= 10 \times 1728 + 0 \times 144 + 7 \times 12 + 11$
$= 17,280 + 0 + 84 + 11 = 17,375$

15. $14411_{\text{five}} = 1 \times 5^4 + 4 \times 5^3 + 4 \times 5^2 + 1 \times 5^1 + 1$
$= 1 \times 625 + 4 \times 125 + 4 \times 25 + 1 \times 5 + 1$
$= 625 + 500 + 100 + 5 + 1 = 1231$

17. $4000_{\text{five}} = 4 \times 5^3 + 0 \times 5^2 + 0 \times 5^1 + 0$
$= 4 \times 125 + 0 \times 25 + 0 \times 5 + 0$
$= 500 + 0 + 0 + 0 = 500$

19. $463_{\text{seven}} = 4 \times 7^2 + 6 \times 7^1 + 3$
$= 4 \times 49 + 6 \times 7 + 3$
$= 196 + 42 + 3 = 241$

21. The place values for base five are 1, 5, 25, 125, 625, etc.
$$
\begin{array}{r}
1 \\
25\overline{)43} \\
\underline{25} \\
18
\end{array}
\qquad
\begin{array}{r}
3 \\
5\overline{)18} \\
\underline{15} \\
3
\end{array}
$$
$43 = 133_{\text{five}}$

23. The place values for base nine are 1, 9, 81, 729, 6561, etc.
$$
\begin{array}{r}
6 \\
729\overline{)4673} \\
\underline{4374} \\
299
\end{array}
\qquad
\begin{array}{r}
3 \\
81\overline{)299} \\
\underline{243} \\
56
\end{array}
\qquad
\begin{array}{r}
6 \\
9\overline{)56} \\
\underline{54} \\
2
\end{array}
$$
$4673 = 6362_{\text{nine}}$

25. The place values for base two are 1, 2, 4, 8, 16, 32, etc.
$$
\begin{array}{r}
1 \\
16\overline{)17} \\
\underline{16} \\
1
\end{array}
\qquad
\begin{array}{r}
0 \\
8\overline{)1} \\
\underline{0} \\
1
\end{array}
\qquad
\begin{array}{r}
0 \\
4\overline{)1} \\
\underline{0} \\
1
\end{array}
\qquad
\begin{array}{r}
0 \\
2\overline{)1} \\
\underline{0} \\
1
\end{array}
$$
$17 = 10001_{\text{two}}$

27. The place values for base four are 1, 4, 16, 64, 256, etc.
$$
\begin{array}{r}
1 \\
64\overline{)91} \\
\underline{64} \\
27
\end{array}
\qquad
\begin{array}{r}
1 \\
16\overline{)27} \\
\underline{16} \\
11
\end{array}
\qquad
\begin{array}{r}
2 \\
4\overline{)11} \\
\underline{8} \\
3
\end{array}
$$
$91 = 1123_{\text{four}}$

29. The place values for base seven are 1, 7, 49, 343, 2401, etc.

$$\begin{array}{r} 1 \\ 343\overline{)434} \\ \underline{343} \\ 91 \end{array} \qquad \begin{array}{r} 1 \\ 49\overline{)91} \\ \underline{49} \\ 42 \end{array} \qquad \begin{array}{r} 6 \\ 7\overline{)42} \\ \underline{42} \\ 0 \end{array}$$

$434 = 1160_{\text{seven}}$

31.
$$\begin{array}{r} 1 \\ 263_{\text{nine}} \\ + \ 18_{\text{nine}} \\ \hline 282_{\text{nine}} \end{array}$$

33.
$$\begin{array}{r} 1\ 1 \\ 111010_{\text{two}} \\ + \ \ 1101_{\text{two}} \\ \hline 1000111_{\text{two}} \end{array}$$

35.
$$\begin{array}{r} 1\ \ 1 \\ 5A79_{\text{twelve}} \\ + \ B068_{\text{twelve}} \\ \hline 14B25_{\text{twelve}} \end{array}$$

37.
$$\begin{array}{r} {\scriptstyle 0\ 1\ 10\ 0\ \ 1\ 10\ 10} \\ 1\cancel{1}\ \cancel{0}\ \cancel{0}\ \cancel{1}\ \cancel{0}\ \cancel{1}\ \cancel{0}_{\text{two}} \\ - \ \ 1\ 1\ 0\ 0\ 1\ 1_{\text{two}} \\ \hline 1\ 0\ 0\ 1\ 0\ 1\ 1\ 1_{\text{two}} \end{array}$$

39.
$$\begin{array}{r} {\scriptstyle 1\ \ 11} \\ 3\cancel{2}\ \cancel{1}3_{\text{four}} \\ - \ 2123_{\text{four}} \\ \hline 1030_{\text{four}} \end{array}$$

41.
$$\begin{array}{r} {\scriptstyle 2\ 2} \\ 254_{\text{six}} \\ \times \ \ \ 3_{\text{six}} \\ \hline 1250_{\text{six}} \end{array}$$

43.
$$\begin{array}{r} 151_{\text{eight}} \quad \text{remainder } 3_{\text{eight}} \\ 7_{\text{eight}}\overline{)1342_{\text{eight}}} \\ \underline{7} \\ 44 \\ \underline{43} \\ 12 \\ \underline{7} \\ 3 \end{array}$$

5 The Real Number System

Exercise Set 5-1

1. 1; 2; 4; 8; 16

3. 1; 2; 3; 6; 7; 9; 14; 18; 21; 42; 63; 126

5. 1; 2; 4; 8; 16; 32

7. 1; 3; 9

9. 1; 2; 3; 4; 6; 8; 12; 16; 24; 32; 48; 96

11. 1; 17

13. 1; 2; 4; 8; 16; 32; 64

15. 1; 3; 5; 7; 15; 21; 35; 105

17. 1; 2; 7; 14; 49; 98

19. 1; 71

21. Five multiples of 3 are $3 \times 5 = 15$, $3 \times 6 = 18$, $3 \times 7 = 21$, $3 \times 8 = 24$, and $3 \times 9 = 27$.

23. Five multiples of 10 are $10 \times 2 = 20$, $10 \times 3 = 30$, $10 \times 4 = 40$, $10 \times 5 = 50$, and $10 \times 6 = 60$.

25. Five multiples of 15 are $15 \times 2 = 30$, $15 \times 3 = 45$, $15 \times 4 = 60$, $15 \times 5 = 75$, and $15 \times 6 = 90$.

27. Five multiples of 17 are $17 \times 2 = 34$, $17 \times 3 = 51$, $17 \times 4 = 68$, $17 \times 5 = 85$, and $17 \times 6 = 102$.

29. Five multiples of 1 are $1 \times 2 = 2$, $1 \times 3 = 3$, $1 \times 4 = 4$, $1 \times 5 = 5$, and $1 \times 6 = 6$.

31.
```
   16
  /  \
 2 × 8
 |  | \
 2 × 2 × 4
 |  |  | \
 2 × 2 × 2 × 2 or 2⁴
```
$2 \times 2 \times 2 \times 2$ or 2^4

33.
```
        1296
       /   \
     2 × 648
     |   |  \
    2 × 2 × 324
    |   |   | \
   2 × 2 × 2 × 162
   |   |   |  | \
  2 × 2 × 2 × 2 × 81
  |   |   |   |  | \
 2 × 2 × 2 × 2 × 3 × 27
 |   |   |   |  |  | \
2 × 2 × 2 × 2 × 3 × 3 × 9
|   |   |   |  |  |  | \
2 × 2 × 2 × 2 × 3 × 3 × 3 × 3
```
$2 \times 2 \times 2 \times 2 \times 3 \times 3 \times 3 \times 3$ or $2^4 \times 3^4$

35. 17 is prime, so no prime factorization exists.

37.
```
   50
  /  \
 2 × 25
 |  | \
 2 × 5 × 5
```
$2 \times 5 \times 5$ or 2×5^2

39.
```
         128
        /  \
       2 × 64
       |  | \
      2 × 2 × 32
      |  |  | \
     2 × 2 × 2 × 16
     |  |  |  | \
    2 × 2 × 2 × 2 × 8
    |  |  |  |  | \
   2 × 2 × 2 × 2 × 2 × 4
   |  |  |  |  |  | \
  2 × 2 × 2 × 2 × 2 × 2 × 2
```
$2 \times 2 \times 2 \times 2 \times 2 \times 2 \times 2$ or 2^7

41.
```
       300
      /  \ `
    2 × 150
    |  |  \
   2 × 2 × 75
   |  |  | \
  2 × 2 × 3 × 25
  |  |  |  | \
 2 × 2 × 3 × 5 × 5
```
$2 \times 2 \times 3 \times 5 \times 5$ or $2^2 \times 3 \times 5^2$

43.
```
     475
    /  \
  5 × 95
  |  | \
 5 × 5 × 19
```
$5 \times 5 \times 19$ or $5^2 \times 19$

45.

$$448$$
$$/ \ \backslash$$
$$7 \times 64$$
$$|\ \ |\ \ \backslash$$
$$7 \times 2 \times 32$$
$$|\ \ |\ \ |\ \ \backslash$$
$$7 \times 2 \times 2 \times 16$$
$$|\ \ |\ \ |\ \ |\ \ \backslash$$
$$7 \times 2 \times 2 \times 2 \times 8$$
$$|\ \ |\ \ |\ \ |\ \ |\ \ \backslash$$
$$7 \times 2 \times 2 \times 2 \times 2 \times 4$$
$$|\ \ |\ \ |\ \ |\ \ |\ \ |\ \ \backslash$$
$$7 \times 2 \times 2 \times 2 \times 2 \times 2 \times 2 \text{ or } 7 \times 2^6$$

47.

$$247$$
$$/ \ \backslash$$
$$13 \times 19$$

49.

$$750$$
$$/ \ \backslash$$
$$2 \times 375$$
$$|\ \ |\ \ \backslash$$
$$2 \times 3 \times 125$$
$$|\ \ |\ \ |\ \ \backslash$$
$$2 \times 3 \times 5 \times 25$$
$$|\ \ |\ \ |\ \ |\ \ \backslash$$
$$2 \times 3 \times 5 \times 5 \times 5 \text{ or } 2 \times 3 \times 5^3$$

51. **Step 1** 3 is prime.

$$9 = 3 \times 3 \text{ or } 3^2$$

Step 2 One factor of 3 is common to 3 and 9.

Step 3 The GCF is 3.

53. **Step 1** 7 is prime.

$$10 = 2 \times 5$$

Step 2 There are no common factors other than 1.

Step 3 The GCF is 1.

55. **Step 1** $12 = 2 \times 2 \times 3$ or $2^2 \times 3$

$$24 = 2 \times 2 \times 2 \times 3 \text{ or } 2^3 \times 3$$
$$48 = 2 \times 2 \times 2 \times 2 \times 3 \text{ or } 2^4 \times 3$$

Step 2 The factors 2 and 3 are common to 12, 24, and 48.

Step 3 The GCF is $2^2 \times 3 = 12$.

57. **Step 1** $75 = 3 \times 5^2$

$$100 = 2^2 \times 5^2$$

Step 2 The factor 5 is common to 75 and 100.

Step 3 The GCF is $5^2 = 25$.

59. **Step 1** $100 = 2^2 \times 5^2$

$$225 = 3^2 \times 5^2$$
$$350 = 2 \times 7 \times 5^2$$

Step 2 The factor 5 is common to 100, 225, and 350.

Step 3 The GCF is $5^2 = 25$.

61. **Step 1** 5 is prime.

$$10 = 2 \times 5$$

Step 2 The different prime factors are 2 and 5.

Step 3 The LCM is $2 \times 5 = 10$.

63. **Step 1** $50 = 2 \times 5^2$

$$75 = 3 \times 5^2$$

Step 2 The different prime factors are 2, 3, and 5.

Step 3 The LCM is $2 \times 3 \times 5^2 = 150$.

65. **Step 1** $70 = 2 \times 5 \times 7$

$$90 = 2 \times 3^2 \times 5$$

Step 2 The different prime factors are 2, 3, 5, and 7.

Step 3 The LCM is $2 \times 3^2 \times 5 \times 7 = 630$.

67. **Step 1** $4 = 2^2$

7 is prime.
11 is prime.

Step 2 The different prime factors are 2, 7, and 11.

Step 3 The LCM is $2^2 \times 7 \times 11 = 308$.

69. **Step 1** $12 = 2^2 \times 3$

$$18 = 2 \times 3^2$$
$$36 = 2^2 \times 3^2$$

Step 2 The different prime factors are 2 and 3.

Step 3 The LCM is $2^2 \times 3^2 = 36$.

71. Find the LCM of 6 and 8.

$6 = 2 \times 3$

$8 = 2^3$

The LCM is $2^3 \times 3 = 24$. After 24 stops the buses will return to station one at the same time. 24 stops takes 24×5 minutes = 120 minutes, or 2 hours. 2 hours after 10:00 A.M. is noon.

73. Find the GCF of 24 and 18.

$24 = 2^3 \times 3$

$18 = 2 \times 3^2$

The GCF is $2 \times 3 = 6$. Each group of items will consist of 6 items. There will be $24 \div 6 = 4$ groups of six pencils in each group, and there will be $18 \div 6 = 3$ groups of six pictures in each group.

75. Find the LCM of 45 and 30.

$45 = 3^2 \times 5$

$30 = 2 \times 3 \times 5$

The LCM is $2 \times 3^2 \times 5 = 90$. A person can use both clubs for free on the same day after 90 days.

77. All the numbers in the infinite list 1, 2, 3, 4, ...

79. It has exactly one factor = itself = 1.

81. The factorization into primes is unique, if a certain order (say, ascending) is agreed on. That is, for a given composite number, there is exactly one set of prime factors.

83. Yes, but only for the trivial case when factor = the number (call it n) = multiple. At other times, no, because a factor would be smaller than n, whereas a multiple would be bigger than n.

85. One; it is 2.

87.

$3 + 1 = 4$	$7 + 7 = 14$
$3 + 3 = 6$	$3 + 13 = 16$
$3 + 5 = 8$	$5 + 13 = 18$
$5 + 5 = 10$	$3 + 17 = 20$
$5 + 7 = 12$	

Exercise Set 5-2

1. $-6 + 5 = -1$

3. $16 + (-7) = 9$

5. $(-8) + (-3) = -11$

7. $-3 + (-9) = -12$

9. $-3 + (-4) + (-6) = -7 + (-6) = -13$

11. $8 - (-6) = 8 + 6 = 14$

13. $6 - 11 = 6 + (-11) = -5$

15. $-3 - (-4) = -3 + 4 = 1$

17. $-12 - (-7) = -12 + 7 = -5$

19. $-20 - 50 = -20 + (-50) = -70$

21. $(5)(9) = 45$

23. $(-3)(8) = -24$

25. $4(-9) = -36$

27. $(-3)(-14) = 42$

29. $(-9)(0) = 0$

31. $64 \div 8 = 8$

33. $-25 \div 5 = -5$

35. $32 \div (-8) = -4$

37. $-14 \div (-2) = 7$

39. $-90 \div (-90) = 1$

41. $0 \div 16 = 0$

43. $-42 \div 6 + 7 = -7 + 7 = 0$

45. $5^3 - 2 \cdot 7 = 125 - 2 \cdot 7 = 125 - 14 = 111$

47. $9 \cdot 9 - 5 \cdot 6 = 81 - 30 = 51$

49. $3^3 + 5^2 - 2^4 = 27 + 25 - 16 = 52 - 16 = 36$

51. $-3[6 + (-10) - (-2)] = -3[-4 - (-2)]$
$= -3(-4 + 2)$
$= -3(-2)$
$= 6$

53. $376 - 14 \cdot 3^4 = 376 - 14 \cdot 81$
$= 376 - 1134$
$= 376 + (-1134)$
$= -758$

55. $256 - 4^3 \cdot 5 + (8 \cdot 4 - 6 \cdot 4)$
$= 256 - 4^3 \cdot 5 + (32 - 24)$
$= 256 - 4^3 \cdot 5 + [32 + (-24)]$
$= 256 - 4^3 \cdot 5 + 8$
$= 256 - 64 \cdot 5 + 8$
$= 256 - 320 + 8$
$= 256 + (-320) + 8$
$= -64 + 8$
$= -56$

57. $-56 \div 8 - \{3 \times [-10 - (4 \times 3)]\}$
$= -56 \div 8 - \{3 \times [-10 - 12]\}$
$= -56 \div 8 - \{3 \times [-10 + (-12)]\}$
$= -56 \div 8 - \{3 \times [-22]\}$
$= -56 \div 8 - \{-66\}$
$= -7 - \{-66\}$
$= -7 + 66$
$= 59$

59. $32 - \{-16 + 5[25 + 9^2 + (8 - 6)]\}$
$= 32 - \{-16 + 5[25 + 9^2 + 2]\}$
$= 32 - \{-16 + 5[25 + 81 + 2]\}$
$= 32 - \{-16 + 5[106 + 2]\}$
$= 32 - \{-16 + 5[108]\}$
$= 32 - \{-16 + 540\}$
$= 32 - 524$
$= 32 + (-524)$
$= -492$

61. $16 < 22$

63. $-5 > -10$

65. $0 > -3$

67. $-9 < +8$

69. $-10 < -7$

71. $|-8| = 8$

73. $|+10| = 10$

75. The opposite of -8 is 8.

77. The opposite of $+10$ is -10.

79. The opposite of 0 is 0.

81. $\$867 + (\$83 + \$562 + \$37 + \$43) - (\$74 + \$86 + \$252) = \$867 + \$725 - \$412$
$$= \$1180$$

83. The number of chicks hatched for the week was
$382 + 494 + 327 + 778 + 256 + 641 = 2878$.
The number of chicks sold for the week was
$105 + 850 + 416 + 237 + 192 + 965 = 2765$.
The number of chicks left at the end of the week was $1286 + 2878 - 2765 = 4164 - 2765 = 1399$.

85. (1) $12,345 - 10,054 = 12,345 + (-10,054)$
$$= 2291$$

(2) $96,686 - 86,409 = 96,686 + (-86,409)$
$$= 10,277$$

(3) $156,860 - 141,449 = 156,860 + (-141,449)$
$$= 15,411$$

(4) $156,284 - 144,514 = 156,284 + (-144,514)$
$$= 11,770$$

(5) $113,443 - 104,925 = 113,443 + (-104,925)$
$$= 8518$$

87. $18 - 10 = 8$ inches

89. A negative number is always less than zero and always to the left of zero on the number line. The opposite of a number may or may not be negative. Example: The respective opposites of 22 and −51 are −22 and 51.

91. (a) To indicate the operation of subtraction

    (b) To indicate position to the left of zero on the number line

93. The set of integers includes all the whole numbers, as well as the integers $-1, -2, -3, -4, \ldots$

95. (a) $-(-3) = 3$

    (b) $-[-(-5)] = -[5] = -5$

    (c) $-[-(-9)] = -[9] = -9$

    (d) $-[-(+10)] = -[-10] = 10$

    (e) $-(-0) = 0$

Exercise Set 5-3

1. The GCF of 7 and 42 is 7.
$$\frac{7 \div 7}{42 \div 7} = \frac{1}{6}$$

3. The GCF of 42 and 60 is 6.
$$\frac{42 \div 6}{60 \div 6} = \frac{7}{10}$$

5. The GCF of 30 and 36 is 6.
$$\frac{30 \div 6}{36 \div 6} = \frac{5}{6}$$

7. The GCF of 91 and 104 is 13.
$$\frac{91 \div 13}{104 \div 13} = \frac{7}{8}$$

9. The GCF of 420 and 756 is 84.
$$\frac{420 \div 84}{756 \div 84} = \frac{5}{9}$$

11. $\dfrac{5}{16} = \dfrac{5 \cdot 3}{16 \cdot 3} = \dfrac{15}{48}$

13. $\dfrac{19}{24} = \dfrac{19 \cdot 2}{24 \cdot 2} = \dfrac{38}{48}$

15. $\dfrac{7}{9} = \dfrac{7 \cdot 5}{9 \cdot 5} = \dfrac{35}{45}$

17. $\dfrac{11}{16} = \dfrac{11 \cdot 5}{16 \cdot 5} = \dfrac{55}{80}$

19. $\dfrac{1}{5} = \dfrac{1 \cdot 6}{5 \cdot 6} = \dfrac{6}{30}$

21. $\dfrac{-5}{6} + \dfrac{2}{3} = \dfrac{-5}{6} + \dfrac{2}{3} \cdot \dfrac{2}{2} = \dfrac{-5}{6} + \dfrac{4}{6} = -\dfrac{1}{6}$

23. $\dfrac{-11}{12} - \dfrac{5}{8} = \dfrac{-11}{12} \cdot \dfrac{2}{2} - \dfrac{5}{8} \cdot \dfrac{3}{3}$
$$= \dfrac{-22}{24} - \dfrac{15}{24}$$
$$= -\dfrac{37}{24} \text{ or } -\left(1\dfrac{13}{24}\right)$$

25. $\dfrac{-5}{12} \times \dfrac{-7}{10} = \dfrac{\overset{1}{\cancel{5}}}{12} \times \left(-\dfrac{7}{\underset{2}{\cancel{10}}}\right) = \dfrac{1 \cdot 7}{12 \cdot 2} = \dfrac{7}{24}$

27. $\dfrac{7}{9} \div \dfrac{2}{3} = \dfrac{7}{9} \cdot \dfrac{3}{2} = \dfrac{7}{\underset{3}{\cancel{9}}} \cdot \dfrac{\overset{1}{\cancel{3}}}{2} = \dfrac{7 \cdot 1}{3 \cdot 2} = \dfrac{7}{6} \text{ or } 1\dfrac{1}{6}$

29. $\left(\dfrac{7}{16} \div \dfrac{3}{8}\right) \times \dfrac{3}{5} = \left(\dfrac{7}{16} \cdot \dfrac{8}{3}\right) \times \dfrac{3}{5}$
$$= \left(\dfrac{7}{\underset{2}{\cancel{16}}} \cdot \dfrac{\overset{1}{\cancel{8}}}{3}\right) \times \dfrac{3}{5}$$
$$= \left(\dfrac{7 \cdot 1}{2 \cdot 3}\right) \times \dfrac{3}{5}$$
$$= \dfrac{7}{6} \times \dfrac{\overset{1}{\cancel{3}}}{5}$$
$$= \dfrac{7}{\underset{2}{\cancel{6}}} \times \dfrac{3}{5}$$
$$= \dfrac{7 \cdot 1}{2 \cdot 5}$$
$$= \dfrac{7}{10}$$

31. $\dfrac{-11}{22} \times \left(\dfrac{1}{6} \times \dfrac{3}{4}\right) = \dfrac{-11}{22} \times \left(\dfrac{1}{\overset{}{\underset{2}{6}}} \times \dfrac{\overset{1}{3}}{4}\right)$

$\qquad = \dfrac{-11}{22} \times \left(\dfrac{1 \cdot 1}{2 \cdot 4}\right)$

$\qquad = \dfrac{-11}{22} \times \dfrac{1}{8}$

$\qquad = \dfrac{-11 \cdot 1}{22 \cdot 8}$

$\qquad = \dfrac{-11}{176}$

$\qquad = -\dfrac{1 \cdot 11}{16 \cdot 11}$

$\qquad = -\dfrac{1}{16}$

33. $\left(\dfrac{5}{8} + \dfrac{3}{4}\right) \times \dfrac{2}{3} = \left(\dfrac{5}{8} + \dfrac{3}{4} \cdot \dfrac{2}{2}\right) \times \dfrac{2}{3}$

$\qquad = \left(\dfrac{5}{8} + \dfrac{6}{8}\right) \times \dfrac{2}{3}$

$\qquad = \dfrac{11}{8} \times \dfrac{2}{3}$

$\qquad = \dfrac{11}{\overset{}{\underset{4}{8}}} \times \dfrac{\overset{1}{2}}{3}$

$\qquad = \dfrac{11 \cdot 1}{4 \cdot 3}$

$\qquad = \dfrac{11}{12}$

35. $\left(\dfrac{-3}{4}\right) \div \left(\dfrac{-5}{8}\right) = -\dfrac{3}{4} \cdot \left(-\dfrac{8}{5}\right)$

$\qquad = -\dfrac{3}{\overset{}{\underset{1}{4}}} \cdot \left(-\dfrac{\overset{2}{8}}{5}\right)$

$\qquad = -\dfrac{3 \cdot 2}{1 \cdot 5}$

$\qquad = \dfrac{6}{5} \text{ or } 1\dfrac{1}{5}$

37. $\left(\dfrac{9}{14} \div \dfrac{3}{7}\right) \times \dfrac{1}{2} = \left(\dfrac{9}{14} \cdot \dfrac{7}{3}\right) \times \dfrac{1}{2}$

$\qquad = \left(\dfrac{\overset{3}{9}}{\overset{}{\underset{2}{14}}} \cdot \dfrac{\overset{1}{7}}{\overset{}{\underset{1}{3}}}\right) \times \dfrac{1}{2}$

$\qquad = \left(\dfrac{3 \cdot 1}{2 \cdot 1}\right) \times \dfrac{1}{2}$

$\qquad = \dfrac{3}{2} \times \dfrac{1}{2}$

$\qquad = \dfrac{3 \cdot 1}{2 \cdot 2}$

$\qquad = \dfrac{3}{4}$

$\overset{2}{7\overline{)14}} \qquad \overset{3}{3\overline{)9}}$

39. $\left(\dfrac{9}{10} - \dfrac{2}{3}\right) \times \dfrac{5}{6} = \left(\dfrac{9}{10} \cdot \dfrac{3}{3} - \dfrac{2}{3} \cdot \dfrac{10}{10}\right) \times \dfrac{5}{6}$

$\qquad = \left(\dfrac{27}{30} - \dfrac{20}{30}\right) \times \dfrac{5}{6}$

$\qquad = \dfrac{7}{30} \times \dfrac{5}{6}$

$\qquad = \dfrac{7}{\overset{}{\underset{6}{30}}} \times \dfrac{\overset{1}{5}}{6}$

$\qquad = \dfrac{7 \cdot 1}{6 \cdot 6}$

$\qquad = \dfrac{7}{36}$

41. $\begin{array}{r} 0.2 \\ 5\overline{)1.0} \\ \underline{-1\,0} \\ 0 \end{array}$

43. $\begin{array}{r} 0.66... \text{ or } 0.\overline{6} \\ 3\overline{)2.00} \\ \underline{-1\,8} \\ 20 \\ \underline{-18} \\ 2 \end{array}$

45. $\begin{array}{r} 2.25 \\ 4\overline{)9.00} \\ \underline{-8} \\ 1\,0 \\ \underline{-8} \\ 20 \\ \underline{-20} \\ 0 \end{array}$

47.
$$
\begin{array}{r}
0.3055\ldots \text{ or } 0.30\overline{5} \\
36\,)\overline{11.0000} \\
-108 \\
\hline
20 \\
-0 \\
\hline
200 \\
-180 \\
\hline
200 \\
-180 \\
\hline
20
\end{array}
$$

49.
$$
\begin{array}{r}
0.75 \\
4\,)\overline{3.00} \\
-2\,8 \\
\hline
20 \\
-20 \\
\hline
0
\end{array}
$$

51.
$$
\begin{array}{r}
0.9411764705882352 \\
51\,)\overline{48.0000000000000000} \\
-45\,9 \\
\hline
2\,10 \\
-2\,04 \\
\hline
60 \\
-51 \\
\hline
90 \\
-51 \\
\hline
390 \\
-357 \\
\hline
330 \\
-306 \\
\hline
240 \\
-204 \\
\hline
360 \\
-357 \\
\hline
30 \\
-0 \\
\hline
300 \\
-255 \\
\hline
450 \\
-408 \\
\hline
420 \\
-408 \\
\hline
120 \\
-102 \\
\hline
180 \\
-153 \\
\hline
270 \\
-255 \\
\hline
150 \\
-102 \\
\hline
48
\end{array}
$$

53. $0.875 = \dfrac{875}{1000} = \dfrac{7}{8}$

55. **Step 1** Let $n = 0.\overline{54}$
then $100n = 54.\overline{54}$

Step 2 $\begin{aligned} 100n &= 54.\overline{54} \\ -\ n &=\ \ 0.\overline{54} \\ \hline 99n &= 54 \end{aligned}$

Step 3 $\dfrac{99n}{99} = \dfrac{54}{99}$

$n = \dfrac{54}{99} = \dfrac{6}{11}$

57. $0.375 = \dfrac{375}{1000} = \dfrac{3}{8}$

59. $285 \div \dfrac{3}{5} = \dfrac{285}{1} \cdot \dfrac{5}{3} = \dfrac{\overset{95}{\cancel{285}}}{1} \cdot \dfrac{5}{\underset{1}{\cancel{3}}} = \dfrac{95 \cdot 5}{1 \cdot 1} = \dfrac{475}{1}$ or 475

The conference is 475 miles away.

61. $\dfrac{1}{2}\left(\dfrac{2}{7}\right) = \dfrac{1}{2} \cdot \dfrac{2}{7} = \dfrac{1}{\underset{1}{\cancel{2}}} \cdot \dfrac{\overset{1}{\cancel{2}}}{7} = \dfrac{1 \cdot 1}{1 \cdot 7} = \dfrac{1}{7}$

$\dfrac{1}{7}$ of the budget is spent on television advertisement.

63. $\left(2\dfrac{3}{8}\right)(80) = \left(\dfrac{19}{8}\right)\left(\dfrac{80}{1}\right)$

$= \dfrac{19}{\underset{1}{\cancel{8}}} \cdot \dfrac{\overset{10}{\cancel{80}}}{1}$

$= \dfrac{19 \cdot 10}{1 \cdot 1}$

$= \dfrac{190}{1}$ or 190

The cities are 190 miles apart.

65. $\dfrac{3}{4} \div 10 = \dfrac{3}{4} \div \dfrac{10}{1} = \dfrac{3}{4} \cdot \dfrac{1}{10} = \dfrac{3 \cdot 1}{4 \cdot 10} = \dfrac{3}{40}$

$$
\begin{array}{r}
0.075 \\
40\,)\overline{3.000} \\
-2\,80 \\
\hline
200 \\
-200 \\
\hline
0
\end{array}
$$

Each piece will be $\dfrac{3}{40}$ meter or 0.075 meter or 7.5 cm long.

Exercise Set 5-6

1. $3^5 = 3 \cdot 3 \cdot 3 \cdot 3 \cdot 3 = 243$

3. $8^0 = 1$

5. $(-5)^0 = 1$

7. $3^{-5} = \dfrac{1}{3^5} = \dfrac{1}{3 \cdot 3 \cdot 3 \cdot 3 \cdot 3} = \dfrac{1}{243}$

9. $2^{-6} = \dfrac{1}{2^6} = \dfrac{1}{2 \cdot 2 \cdot 2 \cdot 2 \cdot 2 \cdot 2} = \dfrac{1}{64}$

11. $3^4 \cdot 3^2 = 3^6$ and $3^6 = 729$

13. $4^4 \cdot 4^3 = 4^7$ and $4^7 = 16,384$

15. $\dfrac{3^4}{3^2} = 3^2$ and $3^2 = 9$

17. $\dfrac{2^5}{2^4} = 2^1$ or 2

19. $(5^2)^3 = 5^6$ and $5^6 = 15,625$

21. $3^2 \cdot 3^{-4} = 3^{-2} = \dfrac{1}{3^2}$ and $\dfrac{1}{3^2} = \dfrac{1}{9}$

23. $5^{-3} \cdot 5^{-2} = 5^{-5} = \dfrac{1}{5^5}$ and $\dfrac{1}{5^5} = \dfrac{1}{3125}$

25. $\dfrac{2^5}{2^7} = 2^{-2} = \dfrac{1}{2^2}$ and $\dfrac{1}{2^2} = \dfrac{1}{4}$

27. $\dfrac{4^4}{4^7} = 4^{-3} = \dfrac{1}{4^3}$ and $\dfrac{1}{4^3} = \dfrac{1}{64}$

29. $\dfrac{7^2}{7^3} = 7^{-1} = \dfrac{1}{7}$

31. Move the point 8 places to the left so that it falls between 6 and 2.

$625,000,000 = 6.25 \times 10^8$

33. Move the point 3 places to the right so that it falls between 7 and 3.

$0.0073 = 7.3 \times 10^{-3}$

35. Move the point 11 places to the left so that it falls between the 5 and 2.

$528,000,000,000 = 5.28 \times 10^{11}$

37. Move the point 6 places to the right so that it falls between the 6 and 1.

$0.00000618 = 6.18 \times 10^{-6}$

39. Move the point 4 places to the left so that it falls between the 4 and 3.

$43,200 = 4.32 \times 10^4$

41. Move the point 2 places to the right so that it falls between the 8 and 1.

$0.0814 = 8.14 \times 10^{-2}$

43. Move the point 13 places to the left so that it falls between the 3 and 2.

$32,000,000,000,000 = 3.2 \times 10^{13}$

45. Since the exponent is positive, move the point 4 places to the right.

$5.9 \times 10^4 = 59,000$

47. Since the exponent is negative, move the point 5 places to the left.

$3.75 \times 10^{-5} = 0.0000375$

49. Since the exponent is positive, move the point 3 places to the right.

$2.4 \times 10^3 = 2400$

51. Since the exponent is negative, move the point 6 places to the left.

$3 \times 10^{-6} = 0.000003$

53. Since the exponent is positive, move the point 3 places to the right.

$1 \times 10^3 = 1000$

55. Since the exponent is positive, move the point 9 places to the right.

$8.02 \times 10^9 = 8,020,000,000$

57. Since the exponent is positive, move the point 12 places to the right.

$7 \times 10^{12} = 7,000,000,000,000$

59. $(3 \times 10^4)(2 \times 10^6) = 6 \times 10^{10}$

61. $(6.2 \times 10^{-2})(4.3 \times 10^{-6}) = 26.66 \times 10^{-8}$
$$\approx 2.67 \times 10^{-7}$$

63. $(4 \times 10^4)(2.2 \times 10^{-7}) = 8.8 \times 10^{-3}$

65. $(5 \times 10^{-2})(3 \times 10^{-8}) = 15 \times 10^{-10} = 1.5 \times 10^{-9}$

67. $\dfrac{5 \times 10^4}{2.5 \times 10^2} = \dfrac{5}{2.5} \times \dfrac{10^4}{10^2} = 2 \times 10^{4-2} = 2 \times 10^2$

69. $\dfrac{4.2 \times 10^{-2}}{7 \times 10^{-3}} = \dfrac{4.2}{7} \times \dfrac{10^{-2}}{10^{-3}}$

$= 0.6 \times 10^{-2+3}$

$= 0.6 \times 10^1$

$= 6$ or 6×10^0

71. $\dfrac{6.6 \times 10^3}{1.1 \times 10^5} = \dfrac{6.6}{1.1} \times \dfrac{10^3}{10^5} = 6 \times 10^{3-5} = 6 \times 10^{-2}$

73. $63,000,000 = 6.3 \times 10^7$

$41,000,000 = 4.1 \times 10^7$

$(63,000,000)(41,000,000)$

$= (6.3 \times 10^7)(4.1 \times 10^7)$

$= 25.83 \times 10^{14}$

$\approx 2.58 \times 10^{15}$

75. $600,000,000 = 6 \times 10^8$

$25,000,000 = 2.5 \times 10^7$

$\dfrac{600,000,000}{25,000,000} = \dfrac{6 \times 10^8}{2.5 \times 10^7} = \dfrac{6}{2.5} \times \dfrac{10^8}{10^7} = 2.4 \times 10^1$

77. $0.00000025 = 2.5 \times 10^{-7}$

$0.000004 = 4 \times 10^{-6}$

$(0.00000025)(0.000004) = (2.5 \times 10^{-7})(4 \times 10^{-6})$

$= 10 \times 10^{-13}$

$= 1 \times 10^{-12}$

79. $1.86 \times 10^5 = 186,000$

81. $1.7 \times 10^{-24} = 0.0000000000000000000000017$

83. $5.88 \times 10^{12} = 5,880,000,000,000$

85. $280(5.88 \times 10^{12}) = 1646.4 \times 10^{12}$

$\approx 1.65 \times 10^{15}$ miles

87. $C = 2\pi r$

$\approx 2(3.14)(67,000,000)$

$= 2(3.14)(6.7 \times 10^7)$

$= 42.076 \times 10^7$

$\approx 4.21 \times 10^8$ or about $421,000,000$

or 421 million miles

89. 480 million $= 480,000,000 = 4.8 \times 10^8$

$\dfrac{4.8 \times 10^8}{1.86 \times 10^5} = \dfrac{4.8}{1.86} \times \dfrac{10^8}{10^5} \approx 2.58 \times 10^3 = 2580$ seconds

or $\dfrac{2580}{60} = 43$ minutes

91. Move the decimal point left or right until exactly one nonzero digit lies to the left of the point; let n = number of slots moved. Multiply the new decimal number by

(a) 10^n if the moving was to the left.

(b) 10^{-n} if the moving was to the right.

93. If written in decimal notation

(a) The first number is readily seen to exceed 1 and usually very large.

(b) The second number is readily seen to be less than 1 and usually very small.

95. Answers will vary.

Exercise Set 5-7

1. (a) 5

(b) $13 - 5 = 8$

(c) $a_n = a_1 + (n-1)d$

$a_{12} = 5 + (12-1)(8)$

$= 5(11)(8)$

$= 5 + 88$

$= 93$

(d) $S_n = \dfrac{n(a_1 + a_n)}{2}$

$S_{12} = \dfrac{12(5 + 93)}{2} = 588$

3. (a) 50

(b) $48 - 50 = -2$

(c) $a_n = a_1 + (n-1)d$

$a_{12} = 50 + (12-1)(-2)$

$= 50(11)(-2)$

$= 50 + (-22)$

$= 28$

(d) $S_n = \dfrac{n(a_1 + a_n)}{2} = \dfrac{12(50 + 28)}{2} = 468$

5. (a) $\dfrac{1}{8}$

(b) $\dfrac{35}{24} - \dfrac{19}{24} = \dfrac{16}{24} = \dfrac{2}{3}$

(c) $a_n = a_1 + (n-1)d$

$a_{12} = \dfrac{1}{8} + (12-1)\left(\dfrac{2}{3}\right)$

$= \dfrac{1}{8} + (11)\left(\dfrac{2}{3}\right)$

$= \dfrac{1}{8} + \dfrac{22}{3}$

$= \dfrac{1}{8} \cdot \dfrac{3}{3} + \dfrac{22}{3} \cdot \dfrac{8}{8}$

$= \dfrac{3}{24} + \dfrac{176}{24}$

$= \dfrac{179}{24}$

(d) $S_n = \dfrac{n(a_1 + a_n)}{2}$

$S_{12} = \dfrac{12\left(\frac{1}{8} + \frac{179}{24}\right)}{2}$

$= \dfrac{12\left(\frac{3}{24} + \frac{179}{24}\right)}{2}$

$= \dfrac{12\left(\frac{182}{24}\right)}{2}$

$= \dfrac{\frac{182}{2}}{2}$

$= \dfrac{182}{4}$

$= \dfrac{91}{2}$

7. (a) 0.6

(b) $1.6 - 0.6 = 1$

(c) $a_n = a_1 + (n-1)d$

$a_{12} = 0.6 + (12-1)(1)$

$= 0.6 + (11)(1)$

$= 11.6$

(d) $S_n = \dfrac{n(a_1 + a_n)}{2}$

$S_{12} = \dfrac{12(0.6 + 11.6)}{2} = 73.2$

9. (a) 4

(b) $\dfrac{12}{4} = 3$

(c) $a_n = a_1 r^{n-1}$

$a_{12} = 4(3)^{12-1} = 4(3^{11}) = 708{,}588$

(d) $S_n = \dfrac{a_1(1-r^n)}{1-r}$

$S_{12} = \dfrac{4(1-3^{12})}{1-3} = 1{,}062{,}880$

11. (a) $\dfrac{1}{2}$

(b) $\dfrac{1}{4} \div \dfrac{1}{2} = \dfrac{1}{4} \cdot \dfrac{2}{1} = \dfrac{1}{2}$

(c) $a_n = a_1 r^{n-1}$

$a_{12} = \dfrac{1}{2}\left(\dfrac{1}{2}\right)^{12-1}$

$= \dfrac{1}{2}\left(\dfrac{1}{2}\right)^{11}$

$= \dfrac{1}{4096} \approx 0.000244$

(d) $S_n = \dfrac{a_1(1-r^n)}{1-r}$

$S_{12} = \dfrac{\frac{1}{2}\left[1-\left(\frac{1}{2}\right)^{12}\right]}{1-\frac{1}{2}}$

$= \dfrac{\frac{1}{2}\left[1-\frac{1}{4096}\right]}{\frac{1}{2}}$

$= 1 - \dfrac{1}{4096}$

$= \dfrac{4095}{4096} \approx 0.998$

13. (a) -3

(b) $\dfrac{15}{-3} = -5$

(c) $a_n = a_1 r^{n-1}$

$a_{12} = (-3)(-5)^{12-1}$

$= (-3)(-5)^{11}$

$= 146{,}484{,}375$

(d) $S_n = \dfrac{a_1(1-r^n)}{1-r}$

$S_{12} = \dfrac{-3[1-(-5)^{12}]}{1-(-5)} = 122{,}070{,}312$

15. (a) 1

(b) $\dfrac{3}{1} = 3$

(c) $a_n = a_1 r^{n-1}$

$a_{12} = 1(3)^{12-1} = 3^{11}$ or 177,147

(d) $S_n = \dfrac{a_1(1-r^n)}{1-r}$

$S_{12} = \dfrac{1(1-3^{12})}{1-3} = 265,720$

17. 1; 7; 13; 19; 25

19. $-9; -12; -15; -18; -21$

21. $\dfrac{1}{4}; \dfrac{5}{8}; 1; \dfrac{11}{8}; \dfrac{7}{4}$

23. 12; 24; 48; 96; 192

25. $-5; -\dfrac{5}{4}; -\dfrac{5}{16}; -\dfrac{5}{64}; -\dfrac{5}{256}$

27. $\dfrac{1}{6}; -1; 6; -36; 216$

29. $\dfrac{-15}{5} = -3; \dfrac{45}{-15} = -3; \dfrac{-135}{45} = -3; \dfrac{405}{-135} = -3$

There is a common ratio of successive terms, so the sequence is geometric.

31. $2 - 6 = -4; -2 - 2 = -4; -6 - (-2) = -4;$
$-10 - (-6) = -4$
There is a common difference between successive terms, so the sequence is arithmetic.

33. (a) The machinery depreciates $50 each year, so the common difference $d = -50$. Find a_5 where $a_1 = 1800$.

$a_5 = 1800 + (5-1)(-50) = 1600$

The depreciation during the fifth year is $1600.

(b) First find S_5, the total depreciation for the first five years.

$S_5 = \dfrac{5(1800+1600)}{2} = 8500$

Then the machinery's value at the end of the fifth year is $50,000 - $8500 = $41,500.

35. The labor costs increase by $25 for each additional 10 feet. First find a_9 if $a_1 = 100$ and $d = 25$.

$a_9 = 100 + (9-1)(25) = 300$

Then $S_9 = \dfrac{9(100+300)}{2} = 1800.$

A 90-foot tower costs $1800 to build.

37. After the first year the account has
$a_1 = $500 + (0.05)($500) = $525.$
After the second year the account has
$a_2 = a_1 + (0.05)a_1$
$= $525 + 0.05($525)$
$= $551.25.$

Continue in this manner to find a_{10}.

$a_3 = $551.25 + 0.05($551.25) \approx 578.81
$a_4 = $578.81 + 0.05($578.81) \approx 607.75
$a_5 = $607.75 + 0.05($607.75) \approx 638.14
$a_6 = $638.14 + 0.05($638.14) \approx 670.05
$a_7 = $670.05 + 0.05($670.05) \approx 703.55
$a_8 = $703.55 + 0.05($703.55) \approx 738.73
$a_9 = $738.73 + 0.05($738.73) \approx 775.67
$a_{10} = $775.67 + 0.05($775.67) \approx 814.45

After 10 years, $814.45 will be in the account.

39. A list of numbers related to each other by a specific rule is called a sequence.

41. A list of numbers such that any member, except the first, equals the preceding member multiplied by a fixed quantity is called a geometric sequence.

43. He added 1 + 100 to get 101, 2 + 99 to get 101, and then multiplied 50 × 101 = 5050.

45. $a_1 = \dfrac{3}{10}; \quad r = \dfrac{3}{100} \div \dfrac{3}{10} = \dfrac{\overset{1}{\cancel{3}}}{\underset{10}{\cancel{100}}} \cdot \dfrac{\overset{1}{\cancel{10}}}{\underset{1}{\cancel{3}}} = \dfrac{1}{10}$

$S_n = \dfrac{a_1}{1-r} = \dfrac{\frac{3}{10}}{1 - \frac{1}{10}} = \dfrac{\frac{3}{10}}{\frac{9}{10}} = \dfrac{\overset{1}{\cancel{3}}}{\underset{1}{\cancel{10}}} \cdot \dfrac{\overset{1}{\cancel{10}}}{\underset{3}{\cancel{9}}} = \dfrac{1}{3}$

Review Exercises

1. 1; 2; 3; 6; 13; 26; 39; 78

3. 1; 3; 5; 9; 15; 45

5. 1; 2; 4; 5; 7; 10; 14; 20; 28; 35; 70; 140

7. Five multiples of 4 are $4 \times 2 = 8$, $4 \times 3 = 12$, $4 \times 4 = 16$, $4 \times 5 = 20$, and $4 \times 6 = 24$.

9. Five multiples of 9 are $9 \times 2 = 18$, $9 \times 3 = 27$, $9 \times 4 = 36$, $9 \times 5 = 45$, and $9 \times 6 = 54$.

11. 96
/ \
2×48
| | \
$2 \times 2 \times 24$
| | | \
$2 \times 2 \times 2 \times 12$
| | | | \
$2 \times 2 \times 2 \times 2 \times 6$
| | | | | \
$2 \times 2 \times 2 \times 2 \times 2 \times 3$ or $2^5 \times 3$

13. 250
/ \
2×125
| | \
$2 \times 5 \times 25$
| | | \
$2 \times 5 \times 5 \times 5$ or 2×5^3

15. 600
/ \
2×300
| | \
$2 \times 2 \times 150$
| | | \
$2 \times 2 \times 2 \times 75$
| | | | \
$2 \times 2 \times 2 \times 3 \times 25$
| | | | | \
$2 \times 2 \times 2 \times 3 \times 5 \times 5$ or $2^3 \times 3 \times 5^2$

17. $6 = 2 \times 3$
$10 = 2 \times 5$
The factor 2 is common to 6 and 10 so the GCF is 2. The different prime factors are 2, 3, and 5 so the LCM is 30.

19. $35 = 5 \times 7$
$40 = 2^3 \times 5$
The factor 5 is common to 35 and 40 so the GCF is 5. The different prime factors are 2, 5, and 7 so the LCM is 280.

21. $60 = 2^2 \times 3 \times 5$
$80 = 2^4 \times 5$
$100 = 2^2 \times 5^2$
The factors 2 and 5 are common to 60, 80, and 100 so the GCF is 20. The different prime factors are 2, 3, and 5 so the LCM is 1200.

23. $-6 + 24 = 18$

25. $5(-9) = -45$

27. $6 + (-2) - (-3) = 4 - (-3) = 4 + 3 = 7$

29. $4 \cdot 3 \div (-3) + (-2) = 12 \div (-3) + (-2)$
$= -4 + (-2)$
$= -6$

31. $\{8 \cdot 7^3 - 55[(3+4) - 6]\} + 20$
$= \{8 \cdot 343 - 55[(3+4) - 6]\} + 20$
$= \{8 \cdot 343 - 55[7 - 6]\} + 20$
$= \{8 \cdot 343 - 55(1)\} + 20$
$= \{2744 - 55\} + 20$
$= 2689 + 20$
$= 2709$

33. The GCF of 75 and 95 is 5.
$$\frac{75 \div 5}{95 \div 5} = \frac{15}{19}$$

35. The GCF of 48 and 60 is 12.
$$\frac{48 \div 12}{60 \div 12} = \frac{4}{5}$$

37. $\dfrac{1}{8} + \dfrac{5}{6} = \dfrac{1}{8} \cdot \dfrac{3}{3} + \dfrac{5}{6} \cdot \dfrac{4}{4} = \dfrac{3}{24} + \dfrac{20}{24} = \dfrac{23}{24}$

39. $\dfrac{5}{9} \times \dfrac{3}{7} = \dfrac{5}{\underset{3}{9}} \times \dfrac{\overset{1}{3}}{7} = \dfrac{5 \cdot 1}{3 \cdot 7} = \dfrac{5}{21}$

41. $\dfrac{1}{2} \div \left(\dfrac{2}{3} + \dfrac{3}{4} \right) = \dfrac{1}{2} \div \left(\dfrac{2}{3} \cdot \dfrac{4}{4} + \dfrac{3}{4} \cdot \dfrac{3}{3} \right)$
$= \dfrac{1}{2} \div \left(\dfrac{8}{12} + \dfrac{9}{12} \right)$
$= \dfrac{1}{2} \div \dfrac{17}{12}$
$= \dfrac{1}{\underset{1}{2}} \cdot \dfrac{\overset{6}{12}}{17}$
$= \dfrac{6}{17}$

43. $\dfrac{2}{3} \left(\dfrac{3}{4} + \dfrac{1}{2} - \dfrac{1}{6} \right) = \dfrac{2}{3} \left(\dfrac{3}{4} \cdot \dfrac{3}{3} + \dfrac{1}{2} \cdot \dfrac{6}{6} - \dfrac{1}{6} \cdot \dfrac{2}{2} \right)$
$= \dfrac{2}{3} \left(\dfrac{9}{12} + \dfrac{6}{12} - \dfrac{2}{12} \right)$
$= \dfrac{2}{3} \left(\dfrac{13}{12} \right)$
$= \dfrac{\overset{1}{2}}{3} \cdot \dfrac{13}{\underset{6}{12}}$
$= \dfrac{13}{18}$

45. $-\dfrac{6}{7}\left(\dfrac{1}{2}+2\dfrac{1}{3}\right)=-\dfrac{6}{7}\left(\dfrac{1}{2}+\dfrac{7}{3}\right)$

$=-\dfrac{6}{7}\left(\dfrac{1}{2}\cdot\dfrac{3}{3}+\dfrac{7}{3}\cdot\dfrac{2}{2}\right)$

$=-\dfrac{6}{7}\left(\dfrac{3}{6}+\dfrac{14}{6}\right)$

$=-\dfrac{6}{7}\left(\dfrac{17}{6}\right)$

$=-\dfrac{\overset{1}{\cancel{6}}\,17}{7\,\underset{1}{\cancel{6}}}$

$=-\dfrac{17}{7}$

47. $\dfrac{5}{8}-\dfrac{2}{3}\left(-1+\dfrac{2}{5}\right)=\dfrac{5}{8}-\dfrac{2}{3}\left(-\dfrac{5}{5}+\dfrac{2}{5}\right)$

$=\dfrac{5}{8}-\dfrac{2}{3}\left(-\dfrac{3}{5}\right)$

$=\dfrac{5}{8}+\dfrac{2}{\underset{1}{\cancel{3}}}\cdot\dfrac{\overset{1}{\cancel{3}}}{5}$

$=\dfrac{5}{8}+\dfrac{2}{5}$

$=\dfrac{5}{8}\cdot\dfrac{5}{5}+\dfrac{2}{5}\cdot\dfrac{8}{8}$

$=\dfrac{25}{40}+\dfrac{16}{40}$

$=\dfrac{41}{40}$

49.
```
      0.9
 10) 9.0
     -9 0
     ───
        0
```

51.
```
       0.857142
  7) 6.000000
     -5 6
     ────
        40
       -35
       ────
         50
        -49
        ────
          10
          -7
          ──
          30
         -28
         ───
          20
         -14
         ───
           6
```

53. $0.6875=\dfrac{6875}{10,000}=\dfrac{275}{400}=\dfrac{11}{16}$

55. **Step 1**   Let $n=0.2\overline{5}$
then $10n=2.5\overline{5}$

Step 2 $10n=2.5\overline{5}$
$\underline{\;-\;n=0.2\overline{5}}$
$9n=2.3$

Step 3 $\dfrac{9n}{9}=\dfrac{2.3}{9}$

$n=\dfrac{2.3}{9}\cdot\dfrac{10}{10}=\dfrac{23}{90}$

57. $\sqrt{48}=\sqrt{16\cdot3}=\sqrt{16}\cdot\sqrt{3}=4\sqrt{3}$

59. $\dfrac{7}{\sqrt{5}}=\dfrac{7}{\sqrt{5}}\cdot\dfrac{\sqrt{5}}{\sqrt{5}}=\dfrac{7\sqrt{5}}{\sqrt{25}}=\dfrac{7\sqrt{5}}{5}$

61. $\sqrt{\dfrac{3}{8}}=\dfrac{\sqrt{3}}{\sqrt{8}}$

$=\dfrac{\sqrt{3}}{\sqrt{8}}\cdot\dfrac{\sqrt{8}}{\sqrt{8}}$

$=\dfrac{\sqrt{24}}{\sqrt{64}}$

$=\dfrac{\sqrt{24}}{8}$

$=\dfrac{\sqrt{4\cdot6}}{8}$

$=\dfrac{\sqrt{4}\cdot\sqrt{6}}{8}$

$=\dfrac{2\sqrt{6}}{8}$

$=\dfrac{\sqrt{6}}{4}$

63. $\sqrt{20}+2\sqrt{75}-3\sqrt{5}=\sqrt{4\cdot5}+2\sqrt{25\cdot3}-3\sqrt{5}$

$=\sqrt{4}\cdot\sqrt{5}+2\sqrt{25}\cdot\sqrt{3}-3\sqrt{5}$

$=2\sqrt{5}+2\cdot5\sqrt{3}-3\sqrt{5}$

$=(2-3)\sqrt{5}+10\sqrt{3}$

$=-\sqrt{5}+10\sqrt{3}$

65. $\sqrt{27}\cdot\sqrt{63}=\sqrt{9\cdot3}\cdot\sqrt{9\cdot7}$

$=\sqrt{9}\cdot\sqrt{3}\cdot\sqrt{9}\cdot\sqrt{7}$

$=3\sqrt{3}\cdot3\sqrt{7}$

$=9\sqrt{3\cdot7}$

$=9\sqrt{21}$

67. $\dfrac{\sqrt{20}}{\sqrt{5}}=\sqrt{\dfrac{20}{5}}=\sqrt{4}=2$

69.
$$\sqrt{6}\left(\sqrt{2}+\sqrt{5}\right)=\sqrt{6}\cdot\sqrt{2}+\sqrt{6}\cdot\sqrt{5}$$
$$=\sqrt{6\cdot2}+\sqrt{6\cdot5}$$
$$=\sqrt{12}+\sqrt{30}$$
$$=\sqrt{4\cdot3}+\sqrt{30}$$
$$=2\sqrt{3}+\sqrt{30}$$

71. Rational; real

73. Rational; real

75. Whole; integer; rational; real

77. Inverse for multiplication

79. Closure for addition

81. $4^5=4\cdot4\cdot4\cdot4\cdot4=1024$

83. $(-3)^0=1$

85. $6^{-5}=\dfrac{1}{6^5}=\dfrac{1}{6\cdot6\cdot6\cdot6\cdot6}=\dfrac{1}{7776}$

87. $\dfrac{5^6}{5^2}=5^4=625$

89. $2^3\cdot2^{-5}=2^{-2}=\dfrac{1}{2^2}=\dfrac{1}{4}$

91. Move the point 3 places to the left.
$$3826=3.826\times10^3\approx3.83\times10^3$$

93. Move the point 6 places to the right.
$$0.00000327=3.27\times10^{-6}$$

95. Since the exponent is positive, move the point 11 places to the right.
$$5.8\times10^{11}=580,000,000,000$$

97. Since the exponent is negative, move the point 4 places to the left.
$$6.27\times10^{-4}=0.000627$$

99. $(2\times10^4)(4.6\times10^{-6})=9.2\times10^{-2}$

101. $\dfrac{4.8\times10^4}{2.4\times10^{-6}}=\dfrac{4.8}{2.4}\times\dfrac{10^4}{10^{-6}}=2\times10^{10}$

103. 8; 18; 28; 38; 48; 58
$$a_n=a_1+(n-1)d$$
$$a_9=8+(9-1)(10)=8+8(10)=8+80=88$$
$$S_n=\frac{n(a_1+a_n)}{2}$$
$$S_9=\frac{9(8+88)}{2}=432$$

105. $-13;\ -18;\ -23;\ -28;\ -33;\ -38$
$$a_n=a_1+(n-1)d$$
$$a_9=-13+(9-1)(-5)$$
$$=-13+8(-5)$$
$$=-13+(-40)$$
$$=-53$$
$$S_n=\frac{n(a_1+a_n)}{2}$$
$$S_9=\frac{9[-13+(-53)]}{2}=-297$$

107. 7.5; 15; 30; 60; 120; 240
$$a_n=a_1r^{n-1}$$
$$a_9=7.5(2)^{9-1}=7.5(2^8)=1920$$
$$S_n=\frac{a_1(1-r^n)}{1-r}$$
$$S_9=\frac{7.5(1-2^9)}{1-2}=3832.5$$

109. $\dfrac{1}{9};\dfrac{1}{36};\dfrac{1}{144};\dfrac{1}{576};\dfrac{1}{2304};\dfrac{1}{9216}$
$$a_n=a_1r^{n-1}$$
$$a_9=\left(\frac{1}{9}\right)\left(\frac{1}{4}\right)^{9-1}$$
$$=\left(\frac{1}{9}\right)\left(\frac{1}{4}\right)^{8}$$
$$=\frac{1}{9}\cdot\frac{1}{65,536}$$
$$=\frac{1}{589,824}$$
$$S_n=\frac{a_1(1-r^n)}{1-r}$$
$$S_9=\frac{\frac{1}{9}\left[1-\left(\frac{1}{4}\right)^9\right]}{1-\frac{1}{4}}\approx0.148$$

111. Let $a_1 = 24$, $d = 1$. The year 2000 is the 21st year after 1980, including 1980.

$$a_n = a_1 + (n-1)d$$
$$a_{21} = 24 + (21-1)(1) = 44$$

44 million people were without health insurance.

Chapter Test

1. Integer; rational; real

3. Rational; real

5. Irrational; real

7. Whole; integer; rational; real

9. Integer; rational; real

11. $42 = 2 \times 3 \times 7$

$56 = 2^3 \times 7$

The factors 2 and 7 are common to 42 and 56 so the GCF is 14. The different prime factors are 2, 3, and 7 so the LCM is 168.

13. $150 = 2 \times 3 \times 5^2$

$175 = 5^2 \times 7$

$200 = 2^3 \times 5^2$

The factor 5 is common to 150, 175, and 200 so the GCF is 25. The different prime factors are 2, 3, 5, and 7 so the LCM is 4200.

15. The GCF of 15 and 35 is 5.

$$\frac{15 \div 5}{35 \div 5} = \frac{3}{7}$$

17. The GCF of 112 and 175 is 7.

$$\frac{112 \div 7}{175 \div 7} = \frac{16}{25}$$

19. The GCF of 49 and 70 is 7.

$$\frac{49 \div 7}{70 \div 7} = \frac{7}{10}$$

21. $-5 \cdot (-6) + 3 \cdot 2 = 30 + 6 = 36$

23. $\left(\frac{5}{6} \cdot \frac{3}{4}\right) \div \frac{2}{3} = \frac{15}{24} \div \frac{2}{3} = \frac{15}{\underset{8}{\cancel{24}}} \cdot \frac{\overset{1}{\cancel{3}}}{2} = \frac{15 \cdot 1}{8 \cdot 2} = \frac{15}{16}$

25. $-6 + \frac{1}{4} \div \frac{2}{3} + \sqrt{81} = -6 + \frac{1}{4} \div \frac{2}{3} + 9$

$$= -6 + \frac{1}{4} \cdot \frac{3}{2} + 9$$
$$= -6 + \frac{3}{8} + 9$$
$$= -\frac{48}{8} + \frac{3}{8} + \frac{72}{8}$$
$$= \frac{27}{8}$$

27. $\sqrt{27} + \sqrt{3}(2\sqrt{2} - 1) = \sqrt{27} + \sqrt{3} \cdot 2\sqrt{2} - \sqrt{3}$

$$= \sqrt{9 \cdot 3} + 2\sqrt{6} - \sqrt{3}$$
$$= \sqrt{9} \cdot \sqrt{3} + 2\sqrt{6} - \sqrt{3}$$
$$= 3\sqrt{3} + 2\sqrt{6} - \sqrt{3}$$
$$= (3-1)\sqrt{3} + 2\sqrt{6}$$
$$= 2\sqrt{3} + 2\sqrt{6}$$
$$= 2\left(\sqrt{3} + \sqrt{6}\right)$$

29. $\frac{\sqrt{45}}{\sqrt{5}} = \sqrt{\frac{45}{5}} = \sqrt{9} = 3$

31. $0.875 = \frac{875}{1000} = \frac{35}{40} = \frac{7}{8}$

33. **Step 1** Let $n = 0.\overline{2}$
then $10n = 2.\overline{2}$

Step 2 $10n = 2.\overline{2}$
$\underline{- \quad n = 0.\overline{2}}$
$9n = 2$

Step 3 $\frac{9n}{9} = \frac{2}{9}$
$n = \frac{2}{9}$

35. Commutative for addition.

37. Identity for addition

39. Associative for multiplication

41. $8^4 = 8 \cdot 8 \cdot 8 \cdot 8 = 4096$

43. $6^0 = 1$

45. $5^{-3} \cdot 5^{-2} = 5^{-5} = \frac{1}{5^5} = \frac{1}{3125}$

47. Move the point 3 places to the right.
$0.00236 = 2.36 \times 10^{-3}$

49. Since the exponent is negative, move the point 5 places to the left.

$$-6 \times 10^{-5} = -0.00006$$

51. $\dfrac{2.1 \times 10^9}{7 \times 10^5} = 0.3 \times 10^4 = 3 \times 10^3$

53. $a_1 = \dfrac{3}{4}$

$a_2 = \overset{1}{\underset{2}{\dfrac{3}{4}}}\left(-\dfrac{1}{6}\right) = -\dfrac{1}{8}$

$a_3 = \left(-\dfrac{1}{8}\right)\left(-\dfrac{1}{6}\right) = \dfrac{1}{48}$

$a_4 = \left(\dfrac{1}{48}\right)\left(-\dfrac{1}{6}\right) = -\dfrac{1}{288}$

$a_5 = \left(-\dfrac{1}{288}\right)\left(-\dfrac{1}{6}\right) = \dfrac{1}{1728}$

$a_6 = \left(\dfrac{1}{1728}\right)\left(-\dfrac{1}{6}\right) = -\dfrac{1}{10,368}$

$a_7 = \left(-\dfrac{1}{10,368}\right)\left(-\dfrac{1}{6}\right) = \dfrac{1}{62,208}$

$a_n = a_1 r^{n-1}$

$a_{15} = \left(\dfrac{3}{4}\right)\left(-\dfrac{1}{6}\right)^{15-1} \approx 9.57 \times 10^{-12}$

$S_n = \dfrac{a_1(1-r^n)}{1-r}$

$S_{15} = \dfrac{\left(\dfrac{3}{4}\right)\left[1-\left(-\dfrac{1}{16}\right)^{15}\right]}{1-\left(-\dfrac{1}{6}\right)}$

$= \dfrac{\dfrac{3}{4}\left(1+\dfrac{1}{6^{15}}\right)}{1+\dfrac{1}{6}}$

$= \dfrac{\dfrac{3}{4}\left(\dfrac{6^{15}+1}{6^{15}}\right)}{\dfrac{7}{6}}$

$= \dfrac{3}{4} \cdot \dfrac{6}{7}\left(\dfrac{6^{15}+1}{6^{15}}\right) \approx \dfrac{9}{14}(1) = \dfrac{9}{14}$

55. Bet 1 = $20
Bet 2 = $40
Bet 3 = $80
Bet 4 = $160
Bet 5 = $320
$620 won

6 | Other Mathematical Systems

Exercise Set 6-1

1. $27 \div 12 = 2$ remainder 3; 27 is equivalent to 3.

3. $155 \div 12 = 12$ remainder 11; 155 is equivalent to 11.

5. $18 \div 12 = 1$ remainder 6; 18 is equivalent to 6.

7. $259 \div 12 = 21$ remainder 7; 259 is equivalent to 7.

9. Starting at 12 and counting 5 numbers backward, you will get 7.

11. Starting at 12 and counting 3 numbers backward, you will get 9.

13. Start at 5 and count 9 hours clockwise, ending at 2; $5 + 9 = 2$.

15. Start at 11 and count 11 hours clockwise, ending at 10; $11 + 11 = 10$.

17. Start at 12 and count 3 hours clockwise, ending at 3; $12 + 3 = 3$.

19. Start at 10 and count 20 hours clockwise, ending at 6; $10 + 20 = 6$.

21. Start at 6 and count 5 hours clockwise, then count 12 more hours clockwise, ending at 11; $(6 + 5) + 12 = 11$.

23. Start at 11 and count 8 hours clockwise, then count 3 more hours clockwise, ending at 10; $3 + (11 + 8) = 10$.

25. Starting at 8 and counting 6 numbers backward, you will get 2.

27. Starting at 9 and counting 11 numbers backward, you will get 10.

29. Starting at 12 and counting 6 numbers backward, you will get 6.

31. Starting at 3 and counting 12 numbers backward, you will get 3.

33. Starting at 4 and counting 5 numbers backward, you will get 11.

35. Starting at 3 and counting 11 numbers backward, you will get 4.

37. $3 \times 2 = 6$

39. $8 \times 6 = 48$ and $48 \div 12 = 4$ remainder 0. Hence, $8 \times 6 = 12$.

41. $2 \times 5 = 10$

43. $3 \times 7 = 21$ and $21 \div 12 = 1$ remainder 9. Hence, $3 \times 7 = 9$.

45. $5 \times (6 \times 9) = 5 \times 54 = 270$ and $270 \div 12 = 22$ remainder 6. Hence, $5 \times (6 \times 9) = 6$.

47. $(6 \times 4) \times 7 = 24 \times 7 = 168$ and $168 \div 12 = 14$ remainder 0. Hence, $(6 \times 4) \times 7 = 12$.

49. 12; $12 + 12 = 24$ and $24 \div 12 = 2$ remainder 0.

51. 7; $7 + 5 = 12$

53. 10; $10 + 2 = 12$

55. 5; $5 + 7 = 12$

57. 5; $5 + (-5) = 0$, which is equivalent to 12.

59. None

61. None

63. 1; $1 \times 1 = 1$

65. $2 + (5 + 8) = (2 + 5) + 8$. Both sides reduce to 3.

67. $4 + 12 = 4$

69. $7 \times (5 + 11) = 7 \times 5 + 7 \times 11$. Both sides reduce to 4.

71. $5 + y = 3$
$y = 3 - 5$
Starting at 3 and counting 5 numbers backward, you will get 10.

73. $y - 5 = 8$
$y = 8 + 5$
$8 + 5 = 13$ and $13 \div 12 = 1$ remainder 1. Hence, $y = 1$.

75.

$4 \times 1 = 4$	$4 \times 4 = 4$	$4 \times 7 = 4$	$4 \times 10 = 4$
$2 + y = 1$	$2 + y = 4$	$2 + y = 7$	$2 + y = 10$
$y = 1 - 2$	$y = 4 - 2$	$y = 7 - 2$	$y = 10 - 2$
$y = 11$	$y = 2$	$y = 5$	$y = 8$

77. $6 \times 9 = 54$ and $54 \div 12 = 4$ remainder 6. Hence, $y = 6$.

79. $3 \times (4 + 10) = 3 \times 14 = 42$ and $42 \div 12 = 3$ remainder 6. Hence $y = 6$.

81. It takes 10×3 hours = 30 hours to run the simulation 10 times. $11 + 30 = 41$ and $41 \div 12 = 3$ remainder 5. Hence, it will be 5:00 P.M.

83. 3:11 A.M.

85. 3:42 P.M.

87. 10:18 P.M.

89. 0656

91. 0400

93. 1727

95. 2342

97. $0627 + 3$ hours and 42 minutes
 $= 0927 + 42$ minutes
 $= 1009$ or 10:09 A.M.

99. $1540 - 1$ hour and 4 minutes
 $= 1436$ or 2:36 P.M.

101. It's a nonempty set of elements together with
(a) operation(s) for the elements,
(b) properties of said operations, and
(c) definitions.

103. Given m on the clock, find that positive clock integer r that obeys $m + r = 12$; this $r =$ the sought inverse.

105. Answers will vary.

107. They are all equal to 4 on the clock.

Exercise Set 6-2

1. $4 + 3 = 7$ and $7 \div 5 = 1$ remainder 2; hence $4 + 3 = 2$ (mod 5)

3. $3 + 3 = 6$ and $6 \div 4 = 1$ remainder 2; hence $3 + 3 = 2$ (mod 4).

5. $5 \times 8 = 40$ and $40 \div 9 = 4$ remainder 4; hence $5 \times 8 = 4$ (mod 9).

7. $3 \times 3 = 9$ and $9 \div 4 = 2$ remainder 1; hence $3 \times 3 = 1$ (mod 4).

9. $3 - 8 = -5$, which is equivalent to 4 (mod 9).

11. $2 - 3 = -1$, which is equivalent to 3 (mod 4).

13. $(3 + 5) + 2 = 10$ and $10 \div 7 = 1$ remainder 3; hence $(3 + 5) + 2 = 3$ (mod 7).

15. $2 + (3 + 5) = 10$ and $10 \div 8 = 1$ remainder 2; hence $2 + (3 + 5) = 2$ (mod 8).

17. $4 \times (2 \times 3) = 24$ and $24 \div 6 = 4$; hence $4 \times (2 \times 3) = 0$ (mod 6).

19. $7 \times (3 \times 5) = 105$ and $105 \div 9 = 11$ remainder 6; hence $7 \times (3 \times 5) = 6$ (mod 9).

21. $6 \times (2 - 5) = -18$, which is equivalent to 6 (mod 8).

23. $7 \times (3 - 5) = -14$, which is equivalent to 6 (mod 10).

25. $2 - (3 - 5) = 4$

27. $(4 - 7) - 3 = -6$, which is equivalent to 3 (mod 9).

29. $8 - (2 - 5) = 11$

31. $32 \div 6 = 5$ remainder 2, so $32 = 2$ (mod 6).

33. $135 \div 7 = 19$ remainder 2, so $135 = 2$ (mod 7).

35. $16 \div 9 = 1$ remainder 7, so $16 = 7$ (mod 9).

37. $326 \div 3 = 108$ remainder 2, so $326 = 2$ (mod 3).

39. $987 \div 8 = 123$ remainder 3, so $987 = 3$ (mod 8).

41. 1 (mod 6) = 7, so $y = 4$.

43. 6 (mod 8) = −2, so $y = 3$.

45. 6 (mod 8) = 14, so $y = 2$.

47. 1 (mod 5) = 6, so $y = 2$.

49. 6 (mod 7) = 20, so $y = 5$.

51. Sunday + 30 days = 0 + 30 = 30
$30 \div 7 = 4$ remainder 2
Sunday + 30 days = 2 or Tuesday

53. Tuesday + 45 days = 2 + 45 = 47
$47 \div 7 = 6$ remainder 5
Tuesday + 45 days = 5 or Friday

55. Saturday + 360 days = 6 + 360 = 366
$366 \div 7 = 52$ remainder 2
Saturday + 360 days = 2 or Tuesday

57. $7624138 \div 9 = 847126$ remainder 4, which does not match the checking digit 2. The number is not valid.

59. $13480435 \div 9 = 1497826$ remainder 1, which does not match the checking digit 4. The number is not valid.

61. Let r = any number in the system. The identity, usually denoted by 0, is a number (also in the system) with the property that $r + 0 = r = 0 + r$. In modulo m, the identity = $0 = m$.

63. The numbers 0, 1, 2, 3, ..., $m - 1$ for any mod m are equivalent to the numbers in base ten. Hence, performing operations in base ten will result in an answer that can be converted to mod m by counting on a clock in mod m.

65.

+	0	1	2	3	4	5	6
0	0	1	2	3	4	5	6
1	1	2	3	4	5	6	0
2	2	3	4	5	6	0	1
3	3	4	5	6	0	1	2
4	4	5	6	0	1	2	3
5	5	6	0	1	2	3	4
6	6	0	1	2	3	4	5

×	0	1	2	3	4	5	6
0	0	0	0	0	0	0	0
1	0	1	2	3	4	5	6
2	0	2	4	6	1	3	5
3	0	3	6	2	5	1	4
4	0	4	1	5	2	6	3
5	0	5	3	1	6	4	2
6	0	6	5	4	3	2	1

67. Closure, commutative, identity, inverse, associative, distributive

Exercise Set 6-3

1. $C ? E = C$

3. $E ? E = E$

5. $C ? F = E$

7. $C ? C = D$ and $E ? D = D$, so $E ? (C ? C) = D$.

9. $E ? D = D$ and $D ? E = D$, so $(E ? D) ? E = D$.

11. Yes, the system is closed under the operation because all elements in the body of the table appear in the margins of the table.

13. Yes, the operation is commutative because the elements are symmetrical with respect to the main diagonal.

15. $x = D$

17. □ ∗ □ = ☆

19. ○ ∗ ○ = □

21. $\square * \bigcirc = \triangle$ and $\triangle * \triangle = \triangle$, so
$\triangle * (\square * \bigcirc) = \triangle$.

23. $\bigcirc * \bigcirc = \square$ and $\square * \square = \stackrel{\star}{}$, so
$(\bigcirc * \bigcirc) * \square = \stackrel{\star}{}$.

25. $\triangle * \triangle = \triangle$ and $\triangle * \square = \square$, so
$(\triangle * \triangle) * \square = \square$.

27. Yes, the operation $*$ is commutative because the elements are symmetrical with respect to the main diagonal.

29. Yes

31.

$\cup$	U	A	$\varnothing$
U	U	U	U
A	U	A	A
$\varnothing$	U	A	$\varnothing$

33. Yes, the system is closed under the operation because all elements in the body of the table appear in the margins of the table.

35.

$\cap$	U	A	$\varnothing$
U	U	A	$\varnothing$
A	A	A	$\varnothing$
$\varnothing$	$\varnothing$	$\varnothing$	$\varnothing$

37. Yes, the system is closed under the operation because all elements in the body of the table appear in the margins of the table.

39. A nonempty set of elements (symbols); one or more operations on these elements (symbols); definitions; properties

41. Draw the main diagonal, which is a segment drawn from top left to bottom right. If all entries are symmetrical with respect to this diagonal, then the operation is commutative.

43.

$\vee$	T	F
T	T	T
F	T	F

45.

$\leftrightarrow$	T	F
T	T	F
F	F	T

Closure, commutative, associative, identity, inverse, distributive

Review Exercises

1. $67 \div 5 = 13$ remainder 2, so $67 = 2$ (mod 5).

3. $532 \div 8 = 66$ remainder 4, so $532 = 4$ (mod 8).

5. $22 \div 4 = 5$ remainder 2, so $22 = 2$ (mod 4).

7. $37 \div 10 = 3$ remainder 7, so $37 = 7$ (mod 10).

9. $56 \div 9 = 6$ remainder 2, so $56 = 2$ (mod 9).

11. $173 \div 9 = 19$ remainder 2, so $173 = 2$ (mod 9).

13. $250 \div 10 = 25$, so $250 = 0$ (mod 10).

15. $18 \div 3 = 6$, so $18 = 0$ (mod 3).

17. $4721 \div 8 = 590$ remainder 1, so $4721 = 1$ (mod 8).

19. $1000 \div 12 = 83$ remainder 4, so $1000 = 4$ (mod 12).

21. $5 + 9 = 14$ and $14 \div 11 = 1$ remainder 3; hence $5 + 9 = 3$ (mod 11).

23. $6 \times 6 = 36$ and $36 \div 7 = 5$ remainder 1; hence $6 \times 6 = 1$ (mod 7).

25. $3 - 7 = -4$, which is equivalent to 4 (mod 8).

27. $3 + 2 = 5$ and $5 \div 4 = 1$ remainder 1; hence $3 + 2 = 1$ (mod 4).

29. $6 \times 7 = 42$ and $42 \div 10 = 4$ remainder 2; hence $6 \times 7 = 2$ (mod 10).

31. $3 - 4 = -1$, which is equivalent to 4 (mod 5).

33. $5 \times (3 + 7) = 50$ and $50 \div 8 = 6$ remainder 2; hence $5 \times (3 + 7) = 2$ (mod 8).

35. $3 - (3 - 5) = 5$

37. $5 \times (7 - 9) = -10$, which is equivalent to 2 (mod 12).

39. $4 \times 3 \times 5 = 60$ and $60 \div 9 = 6$ remainder 6; hence $4 \times 3 \times 5 = 6$ (mod 9).

41. 2 (mod 8) = 10, so $y = 4$.

43. 1 (mod 9) = 10, so $y = 3$.

45. 5 (mod 6) = -1, so $y = 1$.

47. 1 (mod 12) = 13, so $y = 11$.

49. $3 \times 5 = 15$ and $15 \div 7 = 2$ remainder 1, so $y = 1$.

51. $i \cdot i = -1$

53. $i \cdot 1 = i$

55. $(-i \cdot i) \cdot (-1) = 1 \cdot (-1) = -1$

57. $i^2 = i \cdot i = -1$

59. $i^{10} = i^2 \cdot i^2 \cdot i^2 \cdot i^2 \cdot i^2$
$= (-1) \cdot (-1) \cdot (-1) \cdot (-1) \cdot (-1)$
$= -1$

61. $y = -i$

63. Yes; $(-i \cdot 1) \cdot i = -i \cdot i = 1$
$-i \cdot (1 \cdot i) = -i \cdot i = 1$

65. Yes; all elements in the body of the table appear in the margins.

67. 1

69. $(-1) \cdot (-1) = 1$, so -1 is the inverse of -1.

Chapter Test

1. $43 \div 6 = 7$ remainder 1, so $43 = 1 \pmod 6$.

3. $15 \div 2 = 7$ remainder 1, so $15 = 1 \pmod 2$.

5. $-6 = 2 \pmod 4$.

7. $8 + 6 = 14$ and $14 \div 10 = 1$ remainder 4; hence $8 + 6 = 4 \pmod{10}$.

9. $4 - 6 = -2$, which is equivalent to $5 \pmod 7$.

11. $5 \times 9 = 45$ and $45 \div 11 = 4$ remainder 1; hence $5 \times 9 = 1 \pmod{11}$.

13. $4 \pmod 6 = 4$, so $y = 2$.

15. $7 \pmod 8 = -1$, so $y = 2$.

17. $0 \pmod{12} = 0$, 12, and 24, so $y = 0$, 4, and 8.

19. $x * z = z$

21. $z * z = s$

23. $(z * x) * x = z * x = z$

25. Yes; the elements are symmetrical with respect to the main diagonal.

7 | Topics in Algebra

Exercise Set 7-1

1. $5x + 12x - 6x = (5 + 12 - 6)x = 11x$

3. $4y - 10y - 12y = (4 - 10 - 12)y = -18y$

5. $3p + 2q - 7 + 6p - 3q - 10$
Combine like terms: $3p + 6p = 9p$
$\qquad\qquad\qquad 2q - 3q = -q$
$\qquad\qquad\qquad -7 - 10 = -17$
The answer is $9p - q - 17$.

7. $8x^2 + 6x - 10 + 15 - 7x + 3x^2$
Combine like terms: $8x^2 + 3x^2 = 11x^2$
$\qquad\qquad\qquad 6x - 7x = -x$
$\qquad\qquad\qquad -10 + 15 = 5$
The answer is $11x^2 - x + 5$.

9. $5(6x - 7) = 5 \cdot 6x - 5 \cdot 7 = 30x - 35$

11. $-4(12x - 10) = -4 \cdot 12x - (-4)(10)$
$\qquad\qquad\qquad = -48x - (-40)$
$\qquad\qquad\qquad = -48x + 40$

13. $3(2x + 6) - 5x + 9 = 6x + 18 - 5x + 9 = x + 27$

15. $-7(3x + 8) - 5x + 6 = -21x - 56 - 5x + 6$
$\qquad\qquad\qquad\qquad = -26x - 50$

17. $3x + 7 + 5(x - 6) = 3x + 7 + 5x - 30 = 8x - 23$

19. $4x - 17 - 5(x - 6) = 4x - 17 - 5x + 30 = -x + 13$

21. $5x - 7 = 5(18) - 7 = 90 - 7 = 83$

23. $-2c + 10 = -2(-3) + 10 = 6 + 10 = 16$

25. $3x^2 + 2x - 6 = 3(5)^2 + 2(5) - 6$
$\qquad\qquad\qquad = 3(25) + 2(5) - 6$
$\qquad\qquad\qquad = 75 + 10 - 6$
$\qquad\qquad\qquad = 79$

27. $9r^2 - 5r - 10 = 9(-7)^2 - 5(-7) - 10$
$\qquad\qquad\qquad = 9(49) - 5(-7) - 10$
$\qquad\qquad\qquad = 441 - (-35) - 10$
$\qquad\qquad\qquad = 466$

29. $5x + 18y + 10 = 5(8) + 18(3) + 10$
$\qquad\qquad\qquad = 40 + 54 + 10$
$\qquad\qquad\qquad = 104$

31. $3x^2 - 2y^2 + 6x = 3(-8)^2 - 2(2)^2 + 6(-8)$
$\qquad\qquad\qquad = 3(64) - 2(4) + 6(-8)$
$\qquad\qquad\qquad = 192 - 8 + (-48)$
$\qquad\qquad\qquad = 136$

33. $13y^2 - 6x^2 + 7y - 6x + 1$
$\qquad = 13(9)^2 - 6(-5)^2 + 7(9) - 6(-5) + 1$
$\qquad = 13(81) - 6(25) + 7(9) - 6(-5) + 1$
$\qquad = 1053 - 150 + 63 - (-30) + 1$
$\qquad = 997$

35. $9x^2 + 7y^2 + 6x + 2y + 5$
$\qquad = 9(1)^2 + 7(5)^2 + 6(1) + 2(5) + 5$
$\qquad = 9(1) + 7(25) + 6(1) + 2(5) + 5$
$\qquad = 9 + 175 + 6 + 10 + 5$
$\qquad = 205$

37. $x + \dfrac{3y}{2} = 8 + \dfrac{3(6)}{2} = 8 + \dfrac{18}{2} = 8 + 9 = 17$

39. $8x^2 - \dfrac{5}{2y} = 8(4)^2 - \dfrac{5}{2(6)}$
$\qquad\qquad\quad = 8(16) - \dfrac{5}{2(6)}$
$\qquad\qquad\quad = 128 - \dfrac{5}{12}$
$\qquad\qquad\quad = 127\dfrac{7}{12}$

41. $A = 2\pi rh = 2(3.14)(6)(10) = 376.8$ sq in.

43. $A = P(1 + RT)$
$\qquad = 5000(1 + 0.07 \cdot 3)$
$\qquad = 5000(1 + 0.21)$
$\qquad = 5000(1.21)$
$\qquad = \$6050$

45. $V = \dfrac{4}{3}\pi r^3$
$\qquad = \dfrac{4}{3}(3.14)(4)^3$
$\qquad = \dfrac{4}{3}(3.14)(64)$
$\qquad \approx 267.95$ mm^3

47. $FV = P(1 + R)^N$
$\qquad = 20,000(1 + 0.06)^8$
$\qquad = 20,000(1.06)^8$
$\qquad \approx \$31,876.96$

49. $SA = 2\pi rh + 2\pi r^2$
$$= 2(3.14)(6)(12) + 2(3.14)(6)^2$$
$$= 2(3.14)(6)(12) + 2(3.14)(36)$$
$$= 452.16 + 226.08$$
$$= 678.24$$

51. $I = prt = 500(0.05)(4) = \$100$

53. $2.2x - 1.08y + 9.6 = 2.2(9) - 1.08(24) + 9.6$
$$= 19.8 - 25.92 + 9.6$$
$$= 3.48$$
About 3 items are defective.

55. $C = \dfrac{5}{9}(F - 32) = \dfrac{5}{9}(50 - 32) = \dfrac{5}{9}(18) = 10°C$

57. $KE = \dfrac{mv^2}{2}$
$$= \dfrac{(30)(200)^2}{2}$$
$$= \dfrac{(30)(40,000)}{2}$$
$$= 600,000 \text{ ergs}$$

59. $R = \dfrac{kl}{d^2} = \dfrac{8 \cdot 100}{(30)^2} = \dfrac{8 \cdot 100}{900} = \dfrac{8}{9}$ or about 0.89

61. $FV = P(1 + R)^N$
$$= 9000(1 + 0.08)^6$$
$$= 9000(1.08)^6$$
$$\approx \$14,281.87$$

63. Variables are used in expressions and formulas.

65. Like terms are terms having the same variables, as well as the same exponents for each variable

67. For any three real numbers a, b, and c, we have $a(b + c) = ab + ac$ and also $(a + b)c = ac + bc$.

69. Formulas are a short/compact "message" for indicating a relationship among certain variables, using algebra rather than words.

Exercise Set 7-2

1. $x + 6 = 32$
$$x + 6 - 6 = 32 - 6$$
$$x = 26$$
The solution set is $\{26\}$.

3. $x - 5 = 54$
$$x - 5 + 5 = 54 + 5$$
$$x = 59$$
The solution set is $\{59\}$.

5. $-5 = x - 2$
$$-5 + 2 = x - 2 + 2$$
$$-3 = x$$
The solution set is $\{-3\}$.

7. $9x = 27$
$$\dfrac{9x}{9} = \dfrac{27}{9}$$
$$x = 3$$
The solution set is $\{3\}$.

9. $-3x = 36$
$$\dfrac{-3x}{-3} = \dfrac{36}{-3}$$
$$x = -12$$
The solution set is $\{-12\}$.

11. $6x + 12 = 48$
$$6x + 12 - 12 = 48 - 12$$
$$6x = 36$$
$$\dfrac{6x}{6} = \dfrac{36}{6}$$
$$x = 6$$
The solution set is $\{6\}$.

13. $-5x + 25 = -55$
$$-5x + 25 - 25 = -55 - 25$$
$$-5x = -80$$
$$\dfrac{-5x}{-5} = \dfrac{-80}{-5}$$
$$x = 16$$
The solution set is $\{16\}$.

15. $2x + 10 = 4x - 30$
$$2x - 4x + 10 = 4x - 4x - 30$$
$$-2x + 10 = -30$$
$$-2x + 10 - 10 = -30 - 10$$
$$-2x = -40$$
$$\dfrac{-2x}{-2} = \dfrac{-40}{-2}$$
$$x = 20$$
The solution set is $\{20\}$.

17. $x = 6x - 55$
$$x - 6x = 6x - 6x - 55$$
$$-5x = -55$$
$$\dfrac{-5x}{-5} = \dfrac{-55}{-5}$$
$$x = 11$$
The solution set is $\{11\}$.

19.
$$-2x = 15 - 5x$$
$$-2x + 5x = 15 - 5x + 5x$$
$$3x = 15$$
$$\frac{3x}{3} = \frac{15}{3}$$
$$x = 5$$
The solution set is $\{5\}$.

21.
$$-6x + 15 = 4x - 25$$
$$-6x - 4x + 15 = 4x - 4x - 25$$
$$-10x + 15 = -25$$
$$-10x + 15 - 15 = -25 - 15$$
$$-10x = -40$$
$$\frac{-10x}{-10} = \frac{-40}{-10}$$
$$x = 4$$
The solution set is $\{4\}$.

23.
$$2(x + 8) = 40$$
$$2x + 16 = 40$$
$$2x + 16 - 16 = 40 - 16$$
$$2x = 24$$
$$\frac{2x}{2} = \frac{24}{2}$$
$$x = 12$$
The solution set is $\{12\}$.

25.
$$3(x + 2) = 26$$
$$3x + 6 = 26$$
$$3x + 6 - 6 = 26 - 6$$
$$3x = 20$$
$$\frac{3x}{3} = \frac{20}{3}$$
$$x = \frac{20}{3} \text{ or } 6\frac{2}{3}$$
The solution set is $\left\{\frac{20}{3}\right\}$ or $\left\{6\frac{2}{3}\right\}$.

27.
$$6 + 3(x - 5) = 2(x - 3)$$
$$6 + 3x - 15 = 2x - 6$$
$$3x - 9 = 2x - 6$$
$$3x - 2x - 9 = 2x - 2x - 6$$
$$x - 9 = -6$$
$$x - 9 + 9 = -6 + 9$$
$$x = 3$$
The solution set is $\{3\}$.

29.
$$6 + 7(x - 3) = 2x + 10$$
$$6 + 7x - 21 = 2x + 10$$
$$7x - 15 = 2x + 10$$
$$7x - 2x - 15 = 2x - 2x + 10$$
$$5x - 15 = 10$$
$$5x - 15 + 15 = 10 + 15$$
$$5x = 25$$
$$\frac{5x}{5} = \frac{25}{5}$$
$$x = 5$$
The solution set is $\{5\}$.

31.
$$12(x - 2) - 10(x + 7) = 14$$
$$12x - 24 - 10x - 70 = 14$$
$$2x - 94 = 14$$
$$2x - 94 + 94 = 14 + 94$$
$$2x = 108$$
$$\frac{2x}{2} = \frac{108}{2}$$
$$x = 54$$
The solution set is $\{54\}$.

33.
$$6(-x + 5) = 2(x + 8) - 10$$
$$-6x + 30 = 2x + 16 - 10$$
$$-6x + 30 = 2x + 6$$
$$-6x - 2x + 30 = 2x - 2x + 6$$
$$-8x + 30 = 6$$
$$-8x + 30 - 30 = 6 - 30$$
$$-8x = -24$$
$$\frac{-8x}{-8} = \frac{-24}{-8}$$
$$x = 3$$
The solution set is $\{3\}$.

35.
$$\frac{5}{6}x = 30$$
$$\frac{6}{5} \cdot \frac{5}{6}x = \frac{6}{5} \cdot 30$$
$$x = 36$$
The solution set is $\{36\}$.

37.
$$\frac{5}{8}x = 40$$
$$\frac{8}{5} \cdot \frac{5}{8}x = \frac{8}{5} \cdot 40$$
$$x = 64$$
The solution set is $\{64\}$.

39. $\dfrac{3}{4}x + 2 = 21$

$\dfrac{3}{4}x + 2 - 2 = 21 - 2$

$\dfrac{3}{4}x = 19$

$\dfrac{4}{3} \cdot \dfrac{3}{4}x = \dfrac{4}{3} \cdot 19$

$x = \dfrac{76}{3}$ or $25\dfrac{1}{3}$

The solution set is $\left\{\dfrac{76}{3}\right\}$ or $\left\{25\dfrac{1}{3}\right\}$.

41. $\dfrac{5x}{6} + \dfrac{x}{3} = 30$

$\dfrac{6}{1} \cdot \dfrac{5x}{6} + \dfrac{6}{1} \cdot \dfrac{x}{3} = \dfrac{6}{1} \cdot 30$

$5x + 2x = 180$

$7x = 180$

$\dfrac{7x}{7} = \dfrac{180}{7}$

$x = \dfrac{180}{7}$ or $25\dfrac{5}{7}$

The solution set is $\left\{\dfrac{180}{7}\right\}$ or $\left\{25\dfrac{5}{7}\right\}$.

43. $\dfrac{7x}{3} + \dfrac{4x}{2} = 28$

$\dfrac{6}{1} \cdot \dfrac{7x}{3} + \dfrac{6}{1} \cdot \dfrac{4x}{2} = \dfrac{6}{1} \cdot 28$

$14x + 12x = 168$

$26x = 168$

$\dfrac{26x}{26} = \dfrac{168}{26}$

$x = \dfrac{84}{13}$ or $6\dfrac{6}{13}$

The solution set is $\left\{\dfrac{84}{13}\right\}$ or $\left\{6\dfrac{6}{13}\right\}$.

45. $\dfrac{7x}{3} + 5 = \dfrac{4x}{8} + 10$

$\dfrac{24}{1} \cdot \dfrac{7x}{3} + \dfrac{24}{1} \cdot 5 = \dfrac{24}{1} \cdot \dfrac{4x}{8} + \dfrac{24}{1} \cdot 10$

$56x + 120 = 12x + 240$

$56x - 12x + 120 = 12x - 12x + 240$

$44x + 120 = 240$

$44x + 120 - 120 = 240 - 120$

$44x = 120$

$\dfrac{44x}{44} = \dfrac{120}{44}$

$x = \dfrac{30}{11}$ or $2\dfrac{8}{11}$

The solution set is $\left\{\dfrac{30}{11}\right\}$ or $\left\{2\dfrac{8}{11}\right\}$.

47. $\dfrac{x}{6} + \dfrac{3x}{2} = \dfrac{4}{5} - \dfrac{2x}{15}$

$\dfrac{30}{1} \cdot \dfrac{x}{6} + \dfrac{30}{1} \cdot \dfrac{3x}{2} = \dfrac{30}{1} \cdot \dfrac{4}{5} - \dfrac{30}{1} \cdot \dfrac{2x}{15}$

$5x + 45x = 24 - 4x$

$50x = 24 - 4x$

$50x + 4x = 24 - 4x + 4x$

$54x = 24$

$\dfrac{54x}{54} = \dfrac{24}{54}$

$x = \dfrac{4}{9}$

The solution set is $\left\{\dfrac{4}{9}\right\}$.

49. $\dfrac{9x}{5} - \dfrac{2x}{7} = \dfrac{3}{5} - \dfrac{6}{7}$

$\dfrac{35}{1} \cdot \dfrac{9x}{5} - \dfrac{35}{1} \cdot \dfrac{2x}{7} = \dfrac{35}{1} \cdot \dfrac{3}{5} - \dfrac{35}{1} \cdot \dfrac{6}{7}$

$63x - 10x = 21 - 30$

$53x = -9$

$\dfrac{53x}{53} = \dfrac{-9}{53}$

$x = -\dfrac{9}{53}$

The solution set is $\left\{-\dfrac{9}{53}\right\}$.

51. $3x + 8 = 2y + 4$

$3x + 8 - 4 = 2y + 4 - 4$

$3x + 4 = 2y$

$\dfrac{3x + 4}{2} = \dfrac{2y}{2}$

$y = \dfrac{3x + 4}{2}$ or $\dfrac{3x}{2} + 2$

53.
$$2 + 5x - 7y = 18$$
$$2 + 5x - 7y + 7y = 7y + 18$$
$$2 + 5x = 7y + 18$$
$$2 - 2 + 5x = 7y + 18 - 2$$
$$5x = 7y + 16$$
$$\frac{5x}{5} = \frac{7y + 16}{5}$$
$$x = \frac{7y + 16}{5}$$

55.
$$7x + 2y = 9$$
$$7x - 7x + 2y = 9 - 7x$$
$$2y = 9 - 7x$$
$$\frac{2y}{2} = \frac{9 - 7x}{2}$$
$$y = \frac{9 - 7x}{2}$$

57. $8x - 5 + 2x = 10x - 10 + 5$
$$10x - 5 = 10x - 5$$
$$-5 = -5$$
Since the resulting equation is true, the solution set is $\{x \mid x \text{ is a real number}\}$.

59. $5(x - 3) + 2 = 5x - 8$
$$5x - 15 + 2 = 5x - 8$$
$$5x - 13 = 5x - 8$$
$$-13 = -8$$
Since the resulting equation is false, the equation has no solution. Hence, the solution set is $\varnothing$.

61.
$$R = \frac{kL}{d^2}$$
$$\frac{d^2}{k} \cdot R = \frac{d^2}{k} \cdot \frac{kL}{d^2}$$
$$\frac{Rd^2}{k} = L$$

63.
$$V = \pi r^2 h$$
$$\frac{V}{\pi r^2} = \frac{\pi r^2 h}{\pi r^2}$$
$$h = \frac{V}{\pi r^2}$$

65.
$$V = lwh$$
$$\frac{V}{lw} = \frac{lwh}{lw}$$
$$h = \frac{V}{lw}$$

67.
$$E = mc^2$$
$$\frac{E}{c^2} = \frac{mc^2}{c^2}$$
$$m = \frac{E}{c^2}$$

69.
$$A = \frac{1}{2}bh$$
$$2 \cdot A = 2 \cdot \frac{1}{2}bh$$
$$2A = bh$$
$$\frac{2A}{b} = \frac{bh}{b}$$
$$h = \frac{2A}{b}$$

71.
$$F = \frac{mv^2}{r}$$
$$r \cdot F = r \cdot \frac{mv^2}{r}$$
$$rF = mv^2$$
$$\frac{rF}{F} = \frac{mv^2}{F}$$
$$r = \frac{mv^2}{F}$$

73. A closed equation contains no variables. An open equation contains at least one variable.

75. Solution set = that set whose individual members are all the solutions of the equation.

77. Multiply both sides by the lowest common denominator of all fractions present. This eliminates all such fractions.

79. Answers will vary.

Exercise Set 7-3

1. $x - 3$

3. $x + 9$

5. $11 - x$

7. $x - 9$

9. $7 - x$

11. $8 \cdot x$

13. $3x + 5$

15. $5x + 3$

17. $2x$

19. $\dfrac{x}{14}$ or $x \div 14$

21. Let x = the number. Then six times a number plus the number is written as $6x + x$.
$$6x + x = 56$$
$$7x = 56$$
$$\dfrac{7x}{7} = \dfrac{56}{7}$$
$$x = 8$$

23. Let x = the number. Then twice a number is written as $2x$ and 32 less than four times the number is written as $4x - 32$.
$$2x = 4x - 32$$
$$2x - 4x = 4x - 4x - 32$$
$$-2x = -32$$
$$\dfrac{-2x}{-2} = \dfrac{-32}{-2}$$
$$x = 16$$

25. Let x = one number. Then the other number is $x - 6$.
$$x + (x - 6) = 28$$
$$2x - 6 = 28$$
$$2x - 6 + 6 = 28 + 6$$
$$2x = 34$$
$$\dfrac{2x}{2} = \dfrac{34}{2}$$
$$x = 17$$
The numbers are $x = 17$ and $x - 6 = 17 - 6 = 11$.

27. Let x = the number. Then twice the number is $2x$ and 24 less than four times the number is $4x - 24$.
$$2x = 4x - 24$$
$$2x - 4x = 4x - o4x - 24$$
$$-2x = -24$$
$$\dfrac{-2x}{-2} = \dfrac{-24}{-2}$$
$$x = 12$$

29. Let x = the number. Then twelve more than a number divided by 2 is written as $\dfrac{x+12}{2}$.
$$\dfrac{x+12}{2} = 20$$
$$2 \cdot \dfrac{x+12}{2} = 2 \cdot 20$$
$$x + 12 = 40$$
$$x + 12 - 12 = 40 - 12$$
$$x = 28$$

31. Let x = the revenue for Coca-Cola and $x + 11$ = the revenue for PepsiCo.
$$x + (x + 11) = 47$$
$$2x + 11 = 47$$
$$2x + 11 - 11 = 47 - 11$$
$$2x = 36$$
$$\dfrac{2x}{2} = \dfrac{36}{2}$$
$$x = 18, \; x + 11 = 29$$
The revenues were \$18 billion for Coca-Cola and \$29 billion for PepsiCo.

33. Let x = Bill's age and $3x$ = Pete's age.
$$x + 3x = 48$$
$$4x = 48$$
$$\dfrac{4x}{4} = \dfrac{48}{4}$$
$$x = 12, \; 3x = 36$$
Bill's age is 12 and Pete's age is 36.

35. Let x = the total amount invested. Then the interest earned at 8% was $0.08(0.5x)$ and the interest earned at 6) was $0.06(0.5x)$.
$$0.08(0.5x) + 0.06(0.5x) = 210$$
$$0.04x + 0.03x = 210$$
$$0.07x = 210$$
$$\dfrac{0.07x}{0.07} = \dfrac{210}{0.07}$$
$$x = 3000$$
The total amount invested was \$3000.

37. Let x = the population of White Oak in 1990. Then $x - 0.06x$ = the population of White Oak in 1998.
$$x + (x - 0.06x) = 16,984$$
$$1.94x = 16,984$$
$$\dfrac{1.94x}{1.94} = \dfrac{16,984}{1.94}$$
$$x \approx 8755$$
The population was 8755 in 1990 and $16,984 - 8755 = 8229$ in 1998.

39. Let x be the length of the shrtest piece. Then the other pieces are $x + 0.5$ and $x + 0.5 + 0.5 = x + 1$.
$$x + (x + 0.5) + (x + 1) = 6$$
$$3x + 1.5 = 6$$
$$3x + 1.5 - 1.5 = 6 - 1.5$$
$$3x = 4.5$$
$$\dfrac{3x}{3} = \dfrac{4.5}{3}$$
$$x = 1.5$$
The pieces are 1.5 feet (or 18 inches), $1.5 + 0.5 = 2$ feet (or 24 inches), and $1.5 + 1 = 2.5$ feet (or 30 inches.

41. Let x = the number of acres on the farm. Then $\frac{1}{2}x$ acres were planted with corn and $\frac{1}{4}x$ acres were planted with beans.

$$\frac{1}{2}x + \frac{1}{4}x + 40 = x$$

$$\frac{4}{1} \cdot \frac{1}{2}x + \frac{4}{1} \cdot \frac{1}{4}x + \frac{4}{1} \cdot 40 = \frac{4}{1} \cdot x$$

$$2x + x + 160 = 4x$$

$$3x + 160 = 4x$$

$$3x - 3x + 160 = 4x - 3x$$

$$160 = x$$

The farm is 160 acres.

43. Let x = the price the person paid for her house.

$$82,000 - x = 0.20x$$

$$82,000 - x + x = 0.20x + x$$

$$82,000 = 1.2x$$

$$\frac{82,000}{1.2} = \frac{1.2x}{1.2}$$

$$x \approx 68,333.33$$

The person paid $68,333.33.

45. Let some letter = the unknown item, and translate appropriate information into an equation. Solve the equation and check the solution.

47. Decreased by; less; less than; subtracted from; diminished

49. Divide by; quotient; out of; ratio

Exercise Set 7-4

1.

3.

5.

7.

9.

11.

$$x + 6 < 11$$

$$x + 6 - 6 < 11 - 6$$

$$x < 5$$

The solution set is $\{x \mid x < 5\}$. The graph of the solution set is

13.

$$x - 7 \le 23$$

$$x - 7 + 7 \le 23 + 7$$

$$x \le 30$$

The solution set is $\{x \mid x \le 30\}$. The graph of the solution set is

15.

$$3x \ge 18$$

$$\frac{3x}{3} \ge \frac{18}{3}$$

$$x \ge 6$$

The solution set is $\{x \mid x \ge 6\}$. The graph of the solution set is

17.

$$7 - x > 42$$

$$7 - 7 - x > 42 - 7$$

$$-x > 35$$

$$\frac{-x}{-1} < \frac{35}{-1}$$

$$x < -35$$

The solution set is $\{x \mid x < -35\}$. The graph of the solution set is

19.

$$\frac{2}{3}x < 18$$

$$\frac{3}{2} \cdot \frac{2}{3}x < \frac{3}{2} \cdot 18$$

$$x < 27$$

The solution set is $\{x \mid x < 27\}$. The graph of the solution set is

21.

$$-10x < 30$$

$$\frac{-10x}{-10} > \frac{30}{-10}$$

$$x > -3$$

The solution set is $\{x \mid x > -3\}$. The graph of the solution set is

23.

$$2x + 8 \ge 32$$

$$2x + 8 - 8 \ge 32 - 8$$

$$2x \ge 24$$

$$\frac{2x}{2} \ge \frac{24}{2}$$

$$x \ge 12$$

The solution set is $\{x \mid x \ge 12\}$. The graph of the solution set is

25.
$$-3x + 12 \le 36$$
$$-3x + 12 - 12 \le 36 - 12$$
$$-3x \le 24$$
$$\frac{-3x}{-3} \ge \frac{24}{-3}$$
$$x \ge -8$$

The solution set is $\{x \mid x \ge -8\}$. The graph of the solution set is

-8

27.
$$6(x - 12) \le 54$$
$$6x - 72 \le 54$$
$$6x - 72 + 72 \le 54 + 72$$
$$6x \le 126$$
$$\frac{6x}{6} \le \frac{126}{6}$$
$$x \le 21$$

The solution set is $\{x \mid x \le 21\}$. The graph of the solution set is

21

29.
$$-5(3 - x) \le 27$$
$$-15 + 5x \le 27$$
$$-15 + 15 + 5x \le 27 + 15$$
$$5x \le 42$$
$$\frac{5x}{5} \le \frac{42}{5}$$
$$x \le 8\frac{2}{5}$$

The solution set is $\left\{x \middle| x \le 8\frac{2}{5}\right\}$. The graph of the solution set is

$8\frac{2}{5}$

31.
$$16 - 5x \ge 22$$
$$16 - 16 - 5x \ge 22 - 16$$
$$-5x \ge 6$$
$$\frac{-5x}{5} \le \frac{6}{-5}$$
$$x \le -\frac{6}{5}$$

The solution set is $\left\{x \middle| x \le -\frac{6}{5}\right\}$. The graph of the solution set is

$-\frac{6}{5}$

33.
$$3(x + 1) - 10 < 2x + 7$$
$$3x + 3 - 10 < 2x + 7$$
$$3x - 7 < 2x + 7$$
$$3x - 7 + 7 < 2x + 7 + 7$$
$$3x < 2x + 14$$
$$x - 2x < 2x - 2x + 14$$
$$x < 14$$

The solution set is $\{x \mid x < 14\}$. The graph of the solution set is

14

35.
$$9 - 5(x + 6) \ge 32$$
$$9 - 5x - 30 \ge 32$$
$$-21 - 5x \ge 32$$
$$-21 + 21 - 5x \ge 32 + 21$$
$$-5x \ge 53$$
$$\frac{-5x}{-5} \le \frac{53}{-5}$$
$$x \le -10\frac{3}{5}$$

The solution set is $\left\{x \middle| x \le -10\frac{3}{5}\right\}$. The graph of the solution set is

$-10\frac{3}{5}$

37.
$$6(2x + 3) \ge 5(2x - 15)$$
$$12x + 18 \ge 10x - 75$$
$$12x - 10x + 18 \ge 10x - 10x - 75$$
$$2x + 18 \ge -75$$
$$2x + 18 - 18 \ge -75 - 18$$
$$2x \ge -93$$
$$\frac{2x}{2} \ge \frac{-93}{2}$$
$$x \ge -46\frac{1}{2}$$

The solution set is $\left\{x \middle| x \ge -46\frac{1}{2}\right\}$. The graph of the solution is

$-46\frac{1}{2}$

39.
$$6x - 7 \ge 5(x - 2) - 17$$
$$6x - 7 \ge 5x - 10 - 17$$
$$6x - 7 \ge 5x - 27$$
$$6x - 5x - 7 \ge 5x - 5x - 27$$
$$x - 7 \ge -27$$
$$x - 7 + 7 \ge -27 + 7$$
$$x \ge -20$$

The solution set is $\{x \mid x \ge -20\}$. The graph of the solution set is

-20

41. Let x represent the amount Mary can spend for an automobile.
$$x + 0.07x + 120 \le 8000$$
$$1.07x + 120 \le 8000$$
$$1.07x + 120 - 120 \le 8000 - 120$$
$$1.07x \le 7880$$
$$\frac{1.07x}{1.07} \le \frac{7880}{1.07}$$
$$x \le 7364.485981$$

Mary can spend at most $7364.48.

43. Let x represent the lowest score Betsy can get on the last exam.

$$\frac{0.78 + 0.68 + x}{3} \geq 0.70$$

$$\frac{1.46 + x}{3} \geq 0.70$$

$$3 \cdot \frac{1.46 + x}{3} \geq 3 \cdot 0.70$$

$$1.46 + x \geq 2.10$$

$$1.46 - 1.46 + x \geq 2.10 - 1.46$$

$$x \geq 0.64$$

Betsy can score 64% or greater on the last exam.

45. $x \geq 58$
$168 \geq 58$
$79 \geq 58$
$75 \geq 58$
$58 \geq 58$

The states that satisfy the inequality are Texas, Florida, Kansas, and Colorado.

47. $26 \leq x < 58$
$26 \leq 26 < 58$
$26 \leq 30 < 58$

The states that satisfy the inequality are Missouri and West Virginia.

49. $11 \leq x \leq 30$
$11 \leq 11 \leq 30$
$11 \leq 14 \leq 30$
$11 \leq 22 \leq 30$
$11 \leq 26 \leq 30$
$11 \leq 30 \leq 30$

The states that satisfy the inequality are Maryland, California, Pennsylvania, Missouri, and West Virginia.

51. The open dot indicates the solution set does *not* include that point. The closed dot indicates inclusion.

53. In the original inequality, replace all occurrences of the unknown by any one member of the proposed solution set. If the inequality now holds true, then the proposed solution set could be correct.

55. Let x represent the final exam grade.

$$0.80 < \frac{0.77 + 0.82 + 0.86 + 2x}{5}$$

$$0.80 < \frac{2.45 + 2x}{5}$$

$$5(0.80) < 5 \cdot \frac{2.45 + 2x}{5}$$

$$4 < 2.45 + 2x$$

$$4 - 2.45 < 2.45 - 2.45 + 2x$$

$$1.55 < 2x$$

$$\frac{1.55}{2} < \frac{2x}{2}$$

$$0.775 < x$$

The student needs a 78% or greater on the final exam.

Exercise Set 7-5

1. $\dfrac{18}{28} = \dfrac{9}{14}$

3. $\dfrac{14}{32} = \dfrac{7}{16}$

5. $\dfrac{12}{15} = \dfrac{4}{5}$

7. $\dfrac{3}{8}$

9. 5 feet = 5 · 12 inches = 60 inches

$$\frac{60}{30} = \frac{2}{1}$$

11. $\dfrac{3}{x} = \dfrac{14}{45}$

$$\frac{3}{x} \bowtie \frac{14}{45}$$

$$3 \cdot 45 = 14x$$

$$135 = 14x$$

$$\frac{135}{14} = \frac{14x}{14}$$

$$x = \frac{135}{14} \text{ or about } 9.64$$

13. $\dfrac{5}{6} = \dfrac{x}{42}$

$$\frac{5}{6} \bowtie \frac{x}{42}$$

$$5 \cdot 42 = 6x$$

$$210 = 6x$$

$$\frac{210}{6} = \frac{6x}{6}$$

$$35 = x$$

15.
$$\frac{x-6}{12} = \frac{1}{3}$$
$$3(x-6) = 12 \cdot 1$$
$$3x - 18 = 12$$
$$3x - 18 + 18 = 12 + 18$$
$$3x = 30$$
$$\frac{3x}{3} = \frac{30}{3}$$
$$x = 10$$

17.
$$\frac{2}{x-3} = \frac{5}{x+8}$$
$$2(x+8) = 5(x-3)$$
$$2x + 16 = 5x - 15$$
$$2x - 5x + 16 = 5x - 5x - 15$$
$$-3x + 16 = -15$$
$$-3x + 16 - 16 = -15 - 16$$
$$-3x = -31$$
$$\frac{-3x}{-3} = \frac{-31}{-3}$$
$$x = \frac{31}{3} \text{ or } 10\frac{1}{3}$$

19.
$$\frac{x-3}{4} = \frac{x+6}{20}$$
$$20(x-3) = 4(x+6)$$
$$20x - 60 = 4x + 24$$
$$20x - 4x - 60 = 4x - 4x + 24$$
$$16x - 60 = 24$$
$$16x - 60 + 60 = 24 + 160$$
$$16x = 84$$
$$\frac{16x}{16} = \frac{84}{16}$$
$$x = \frac{21}{4} \text{ or } 5\frac{1}{4}$$

21. Let x represent the number of vanilla ice cream cones sold in one day.
$$\frac{1 \text{ person buys vanilla}}{5 \text{ people buy ice cream}} = \frac{x}{75 \text{ cones sold}}$$
$$\frac{1}{5} = \frac{x}{75}$$
$$75 = 5x$$
$$\frac{75}{5} = \frac{5x}{5}$$
$$15 = x$$
15 ice cream cones will be vanilla.

23. Let x represent the number of inches of water from melted snow.
$$\frac{1.5 \text{ ft snow}}{2 \text{ in. water}} = \frac{3.5 \text{ ft snow}}{x}$$
$$\frac{1.5}{2} = \frac{3.5}{x}$$
$$1.5x = 2(3.5)$$
$$1.5x = 7$$
$$\frac{1.5x}{1.5} = \frac{7}{1.5}$$
$$x = \frac{14}{3} \text{ or } 4\frac{2}{3}$$
There will be $\frac{14}{3}$ inches or $4\frac{2}{3}$ inches of water.

25. Let x represent the number of gallons of paint needed.
$$\frac{1 \text{ gallon}}{640 \text{ sq ft wall space}} = \frac{x}{2560 \text{ sq ft}}$$
$$\frac{1}{640} = \frac{x}{2560}$$
$$2560 = 640x$$
$$\frac{2560}{640} = \frac{640x}{640}$$
$$4 = x$$
4 gallons of paint will be needed.

27. Let x represent the number of female joggers.
$$\frac{9 \text{ females}}{20 \text{ joggers}} = \frac{x}{220 \text{ joggers}}$$
$$\frac{9}{20} = \frac{x}{220}$$
$$9 \cdot 220 = 20x$$
$$1980 = 20x$$
$$\frac{1980}{20} = \frac{20x}{20}$$
$$99 = x$$
There were 99 female joggers.

29. Let x represent the number of defective calculators in the lot.
$$\frac{2 \text{ defective}}{50 \text{ calculators}} = \frac{x}{1000 \text{ calculators}}$$
$$2 \cdot 1000 = 50x$$
$$2000 = 50x$$
$$\frac{2000}{50} = \frac{50x}{50}$$
$$40 = x$$
40 calculators will be defective.

31. Let x represent the number of sections of Statistics 101, which translates to the number of faculty required to teach the course.

$$\frac{16 \text{ students}}{1 \text{ faculty}} = \frac{128 \text{ students}}{x}$$

$$\frac{16}{1} = \frac{128}{x}$$

$$16x = 128$$

$$\frac{16x}{16} = \frac{128}{16}$$

$$x = 8$$

8 sections should be offered.

33. Let x represent the number of professors to be hired.

$$\frac{1200 \text{ students}}{80 \text{ professors}} = \frac{1500 \text{ students}}{80 + x}$$

$$\frac{1200}{80} = \frac{1500}{80 + x}$$

$$1200(80 + x) = 80(1500)$$

$$96,000 + 1200x = 120,000$$

$$96,000 - 96,000 + 1200x = 120,000 - 96,000$$

$$1200x = 24,000$$

$$\frac{1200x}{1200} = \frac{24,000}{1200}$$

$$x = 20$$

20 professors should be hired.

35. Let x represent the number of cans of paint to purchase.

$$\frac{5 \text{ cans}}{20 \text{ sq ft}} = \frac{x}{13\frac{1}{3} \text{ sq ft}}$$

$$\frac{5}{20} = \frac{x}{\frac{40}{3}}$$

$$5 \cdot \frac{40}{3} = 20x$$

$$\frac{200}{3} = 20x$$

$$\frac{1}{20} \cdot \frac{200}{3} = \frac{1}{20} \cdot 20x$$

$$\frac{10}{3} = x$$

Only $\frac{10}{3}$ or $3\frac{1}{3}$ cans of paint are needed, but 4 cans must be purchased.

37. Let y = amount of interest on $5000

k = 6% interest

x = 4 years

$y = (6\%)(\$5000)(4 \text{ years})$

$y = (0.06)(5000)(4)$

$y = \$1200$

39. Let C = circumference of a circle

d = diameter of a circle

$\pi = 3.14$

$C = \pi d$

$32 \text{ in.} = 3.14d$

$$\frac{32}{3.14} = \frac{3.14d}{3.14}$$

$10.19 \approx d$

The diameter is approximately 10.19 inches.

41. A ratio is a comparison of two quantities using division.

43. A proportion is a statement of equality of two equal ratios.

45. $\dfrac{\$3.39}{10 \text{ pounds}} = \$0.339$ per pound

$\dfrac{\$7.49}{25 \text{ pounds}} = \$0.2996$ per pound

25 pounds for $7.49 is a better buy.

47. $\dfrac{\$1.99}{7 \text{ ounces}} \approx \$0.284$ per ounce

$\dfrac{\$0.50}{1.5 \text{ ounces}} \approx \$0.333$ per ounce

7 ounces for $1.99 is a better buy.

49. $\dfrac{\$2.75}{11.5 \text{ ounces}} \approx \$0.239$ per ounce

$\dfrac{\$7.49}{34.5 \text{ ounces}} \approx \$0.217$ per ounce

34.5 ounces for $7.49 is a better buy.

Exercise Set 7-6

1. $(x + 7)(x + 9) = x \cdot x + x \cdot 9 + 7 \cdot x + 7 \cdot 9$

$\qquad = x^2 + 9x + 7x + 63$

$\qquad = x^2 + 16x + 63$

3. $(x - 7)(x - 10)$

$= x \cdot x + x(-10) + (-7)x + (-7)(-10)$

$= x^2 - 10x - 7x + 70$

$= x^2 - 17x + 70$

5. $(x - 15)(x + 8) = x \cdot x + x \cdot 8 + (-15)x + (-15)(8)$

$\qquad = x^2 + 8x - 15x - 120$

$\qquad = x^2 - 7x - 120$

7. $(2x - 7)(7x - 9)$

$= 2x \cdot 7x + 2x(-9) + (-7)(7x) + (-7)(-9)$

$= 14x^2 - 18x - 49x + 63$

$= 14x^2 - 67x + 63$

9. $(5x+7)(3x-8) = 5x \cdot 3x + 5x(-8) + 7 \cdot 3x + 7(-8)$
$= 15x^2 - 40x + 21x - 56$
$= 15x^2 - 19x - 56$

11. Step 1 $x^2 + 5x + 6 = 0$
Step 2 $(x+2)(x+3) = 0$
Step 3 $x + 2 = 0 \qquad x = 3 = 0$
Step 4 $\quad x + 2 = 0 \qquad\qquad x + 3 = 0$
$\qquad x + 2 - 2 = 0 - 2 \quad\Big| \quad x + 3 - 3 = 0 - 3$
$\qquad\qquad x = -2 \qquad\qquad\quad x = -3$
The solution set is $\{-2, -3\}$.

13. Step 1 $x^2 + x - 12 = 0$
Step 2 $(x+4)(x-3) = 0$
Step 3 $x + 4 = 0 \qquad x - 3 = 0$
Step 4 $\quad x + 4 = 0 \qquad\qquad x - 3 = 0$
$\qquad x + 4 - 4 = 0 \quad\Big| \quad x - 3 + 3 = 0 + 3$
$\qquad\qquad x = -4 \qquad\qquad\quad x = 3$
The solution set is $\{-4, 3\}$.

15. Step 1 $x^2 - 14x - 51 = 0$
Step 2 $(x-17)(x+3) = 0$
Step 3 $x - 17 = 0 \qquad x + 3 = 0$
Step 4 $\quad x - 17 = 0 \qquad\qquad x + 3 = 0$
$\quad x - 17 + 17 = 0 + 17 \,\Big|\, x + 3 - 3 = 0 - 3$
$\qquad\qquad x = 17 \qquad\qquad\quad x = -3$
The solution set is $\{-3, 17\}$.

17. Step 1 $x^2 + 24x - 81 = 0$
Step 2 $(x+27)(x-3) = 0$
Step 3 $x + 27 = 0 \qquad x - 3 = 0$
Step 4 $\quad x + 27 = 0 \qquad\qquad x - 3 = 0$
$\quad x + 27 - 27 = 0 - 27 \,\big|\, x - 3 + 3 = 0 + 3$
$\qquad\qquad x = -27 \qquad\qquad\quad x = 3$
The solution set is $\{-27, 3\}$.

19. Step 1 $x^2 - 8x + 15 = 0$
Step 2 $(x-3)(x-5) = 0$
Step 3 $x - 3 = 0 \qquad x - 5 = 0$
Step 4 $\quad x - 3 = 0 \qquad\qquad x - 5 = 0$
$\quad x - 3 + 3 = 0 + 3 \,\Big|\, x - 5 + 5 = 0 + 5$
$\qquad\qquad x = 3 \qquad\qquad\quad x = 5$
The solution set is $\{3, 5\}$.

21. Step 1 $2x^2 - x - 21 = 0$
Step 2 $(x+3)(2x-7) = 0$
Step 3 $x + 3 = 0 \qquad\qquad 2x - 7 = 0$

Step 4 $\quad x + 3 = 0 \qquad\qquad 2x - 7 = 0$
$\quad x + 3 - 3 = 0 - 3 \,\big|\, 2x - 7 + 7 = 0 + 7$
$\qquad\qquad x = -3 \qquad\qquad\quad 2x = 7$
$\qquad\qquad\qquad\qquad\qquad\quad \dfrac{2x}{2} = \dfrac{7}{2}$
$\qquad\qquad\qquad\qquad\qquad\qquad x = \dfrac{7}{2}$
The solution set is $\left\{-3, \dfrac{7}{2}\right\}$.

23. Step 1 $6x^2 - x - 12 = 0$
Step 2 $(3x+4)(2x-3) = 0$
Step 3 $3x + 4 = 0 \qquad\qquad 2x - 3 = 0$
Step 4 $\quad 3x + 4 = 0 \qquad\qquad 2x - 3 = 0$
$\quad 3x + 4 - 4 = 0 - 4 \,\big|\, 2x - 3 + 3 = 0 + 3$
$\qquad\qquad 3x = -4 \qquad\qquad\quad 2x = 3$
$\qquad\qquad \dfrac{3x}{3} = \dfrac{-4}{3} \qquad\quad \dfrac{2x}{2} = \dfrac{3}{2}$
$\qquad\qquad\quad x = -\dfrac{4}{3} \qquad\qquad x = \dfrac{3}{2}$
The solution set is $\left\{-\dfrac{4}{3}, \dfrac{3}{2}\right\}$.

25. Step 1 $6x^2 - x - 12 = 0$
Step 2 $(3x+4)(2x-3) = 0$
Step 3 $3x + 4 = 0 \qquad\qquad 2x - 3 = 0$
Step 4 $\quad 3x + 4 = 0 \qquad\qquad 2x - 3 = 0$
$\quad 3x + 4 - 4 = 0 - 4 \,\big|\, 2x - 3 + 3 = 0 + 3$
$\qquad\qquad 3x = -4 \qquad\qquad\quad 2x = 3$
$\qquad\qquad \dfrac{3x}{3} = \dfrac{-4}{3} \qquad\quad \dfrac{2x}{2} = \dfrac{3}{2}$
$\qquad\qquad\quad x = -\dfrac{4}{3} \qquad\qquad x = \dfrac{3}{2}$
The solution set is $\left\{-\dfrac{4}{3}, \dfrac{3}{2}\right\}$.

27. Step 1 $5x^2 + 7x - 6 = 0$
Step 2 $(x+2)(5x-3) = 0$
Step 3 $x + 2 = 0 \qquad\qquad 5x - 3 = 0$
Step 4 $\quad x + 2 = 0 \qquad\qquad 5x - 3 = 0$
$\quad x + 2 - 2 = 0 - 2 \,\big|\, 5x - 3 + 3 = 0 + 3$
$\qquad\qquad x = -2 \qquad\qquad\quad 5x = 3$
$\qquad\qquad\qquad\qquad\qquad\quad \dfrac{5x}{5} = \dfrac{3}{5}$
$\qquad\qquad\qquad\qquad\qquad\qquad x = \dfrac{3}{5}$
The solution set is $\left\{-2, \dfrac{3}{5}\right\}$.

29. Step 1 $6x^2 + x - 12 = 0$
Step 2 $(3x - 4)(2x + 3) = 0$
Step 3 $3x - 4 = 0$ $2x + 3 = 0$
Step 4 $3x - 4 = 0$ $2x + 3 = 0$
           $3x - 4 + 4 = 0 + 4$ $2x + 3 - 3 = 0 - 3$
                       $3x = 4$ $2x = -3$
                       $\dfrac{3x}{3} = \dfrac{4}{3}$ $\dfrac{2x}{2} = \dfrac{-3}{2}$
                       $x = \dfrac{4}{3}$ $x = -\dfrac{3}{2}$

The solution set is $\left\{ -\dfrac{3}{2}, \dfrac{4}{3} \right\}$.

31. $3x^2 + x - 1 = 0$
$a = 3$, $b = 1$, and $c = -1$
$$x = \frac{-b \pm \sqrt{b^2 - 4ac}}{2a}$$
$$= \frac{-1 \pm \sqrt{1^2 - 4(3)(-1)}}{2(3)}$$
$$= \frac{-1 \pm \sqrt{1 + 12}}{6}$$
$$= \frac{-1 \pm \sqrt{13}}{6}$$
$$\left\{ \frac{-1 + \sqrt{13}}{6}, \frac{-1 - \sqrt{13}}{6} \right\}$$

33. $2x^2 - 5x - 12 = 0$
$a = 2$, $b = -5$, and $a = -12$
$$x = \frac{-b \pm \sqrt{b^2 - 4ac}}{2a}$$
$$= \frac{-(-5) \pm \sqrt{(-5)^2 - 4(2)(-12)}}{2(2)}$$
$$= \frac{5 \pm \sqrt{25 + 96}}{4}$$
$$= \frac{5 \pm \sqrt{121}}{4}$$
$$= \frac{5 \pm 11}{4}$$
$x = \dfrac{5 + 11}{4}$ or $x = \dfrac{5 - 11}{4}$
$= \dfrac{16}{4}$ $= \dfrac{-6}{4}$
$= 4$ $= -\dfrac{3}{2}$
$\left\{ -\dfrac{3}{2}, 4 \right\}$

35. $3x^2 + 5x + 1 = 0$
$a = 3$, $b = 5$, and $c = 1$
$$x = \frac{-b \pm \sqrt{b^2 - 4ac}}{2a}$$
$$= \frac{-5 \pm \sqrt{5^2 - 4(3)(1)}}{2(3)}$$
$$= \frac{-5 \pm \sqrt{25 - 12}}{6}$$
$$= \frac{-5 \pm \sqrt{13}}{6}$$
$$\left\{ \frac{-5 + \sqrt{13}}{6}, \frac{-5 - \sqrt{13}}{6} \right\}$$

37. $x^2 - 8x - 9 = 0$
$a = 1$, $b = -8$, and $c = -9$
$$x = \frac{-b \pm \sqrt{b^2 - 4ac}}{2a}$$
$$= \frac{-(-8) \pm \sqrt{(-8)^2 - 4(1)(-9)}}{2(1)}$$
$$= \frac{8 \pm \sqrt{64 + 36}}{2}$$
$$= \frac{8 \pm \sqrt{100}}{2}$$
$$= \frac{8 \pm 10}{2}$$
$x = \dfrac{8 + 10}{2}$ or $x = \dfrac{8 - 10}{2}$
$= \dfrac{18}{2}$ $= \dfrac{-2}{2}$
$= 9$ $= -1$
$\{-1, 9\}$

39. $x^2 + 5x - 3 = 0$
$a = 1$, $b = 5$, and $c = -3$
$$x = \frac{-b \pm \sqrt{b^2 - 4ac}}{2a}$$
$$= \frac{-5 \pm \sqrt{5^2 - 4(1)(-3)}}{2(1)}$$
$$= \frac{-5 \pm \sqrt{25 + 12}}{2}$$
$$= \frac{-5 \pm \sqrt{37}}{2}$$
$$\left\{ \frac{-5 + \sqrt{37}}{2}, \frac{-5 - \sqrt{37}}{2} \right\}$$

41. $d = rt + 16t^2$, where
$d = 1296$ feet
$r = 0$ feet per second

$$1296 = 0t + 16t^2$$
$$16t^2 - 1296 = 0$$
$$16(t^2 - 81) = 0$$
$$16(t + 9)(t - 9) = 0$$

$t + 9 = 0 \qquad \mid \qquad t - 9 = 0$
$\quad t = -9 \qquad \mid \qquad \quad t = 9$

It takes 9 seconds.

43. Let x = first integer and $x + 1$ = second integer.
$$x(x + 1) = 156$$
$$x^2 + x = 156$$
$$x^2 + x - 156 = 0$$
$$(x + 13)(x - 12) = 0$$

$x + 13 = 0 \qquad \mid \qquad x - 12 = 0$
$\quad x = -13 \qquad \mid \qquad \quad x = 12$

When $x = -13$, the integers are -13 and -12.
When $x = 12$, the integers are 12 and 13.

45. Let t = side of the plot for cabbage, carrots, and lettuce. $t + 6$ = side of the plot for tomatoes.
$$A = s^2$$
$$116 = t^2 + (t + 6)^2$$
$$116 = t^2 + (t + 6)(t + 6)$$
$$116 = t^2 + t^2 + 12t + 36$$
$$116 = 2t^2 + 12t + 36$$

$$2t^2 + 12t - 80 = 0$$
$$2(t^2 + 6t - 40) = 0$$
$$2(t + 10)(t - 4) = 0$$

$t + 10 = 0 \qquad \mid \qquad t - 4 = 0$
$\quad t = -10 \qquad \mid \qquad \quad t = 4$

The plots are 4 feet on a side and 6 feet on a side.

47. The term $x - 1 = 0$. Division by 0 is undefined.

Review Exercises

1. $6x + 3y - 10 + 2y - 8x + 3$
Combine like terms: $6x - 8x = -2x$
$$3y + 2y = 5y$$
$$-10 + 3 = -7$$
the answer is $-2x + 5y - 7$.

3. $5(x - 6) + 2(x - 3) = 5x - 30 + 2x - 6 = 7x - 36$

5. $2x + 7(x - 3) + 4x = 2x + 7x - 21 + 4x$
$$= 13x - 21$$

7. $2x^2 + 5x - 3 = 2(6)^2 + 5(6) - 3$
$$= 2(36) + 5(6) - 3$$
$$= 72 + 30 - 3$$
$$= 99$$

9. $6(x - 8) - 10 = 6(-2 - 8) - 10$
$$= 6(-10) - 10$$
$$= -60 - 10$$
$$= -70$$

11. $d = rt = 8 \cdot 15 = 120$

13.
$$4x + 8 = -32$$
$$4x + 8 - 8 = -32 - 8$$
$$4x = -40$$
$$\frac{4x}{4} = \frac{-40}{4}$$
$$x = -10$$
The solution set is $\{-10\}$.

15.
$$8x - 3 = 6x + 37$$
$$8x - 6x - 3 = 6x - 6x + 37$$
$$2x - 3 = 37$$
$$2x - 3 + 3 = 37 + 3$$
$$2x = 40$$
$$\frac{2x}{2} = \frac{40}{2}$$
$$x = 20$$
The solution set is $\{20\}$.

17.
$$5(x + 9) = -20$$
$$5x + 45 = -20$$
$$5x + 45 - 45 = -20 - 45$$
$$5x = -65$$
$$\frac{5x}{5} = \frac{-65}{5}$$
$$x = -13$$
The solution set is $\{-13\}$.

19. $6(x + 8) - 4x = 3x - 19$
$$6x + 48 - 4x = 3x - 19$$
$$2x + 48 = 3x - 19$$
$$2x - 2x + 48 = 3x - 2x - 19$$
$$48 = x - 19$$
$$48 + 19 = x - 19 + 19$$
$$67 = x$$
The solution set is $\{67\}$.

21.
$$5x + 3 = 16x + 47$$
$$5x - 16x + 3 = 16x - 16x + 47$$
$$-11x + 3 = 47$$
$$-11x + 3 - 3 = 47 - 3$$
$$-11x = 44$$
$$\frac{-11x}{-11} = \frac{44}{-11}$$
$$x = -4$$
The solution set is $\{-4\}$.

23. $8n - 4$

25. $4n + 3$

27. Let x = the time to drive to the conference. Then $8 - x$ is the time to drive back. The distance each way is the same.
distance = rate × time
distance to the conference
= 40 miles per hour $\cdot$ x hours
distance back from the converence
= 50 miles per hour $\cdot$ $(8 - x)$ hours
$$40x = 50(8 - x)$$
$$40x = 400 - 50x$$
$$40x + 50x = 400 - 50x + 50x$$
$$90x = 400$$
$$\frac{90x}{90} = \frac{400}{90}$$
$$x = \frac{40}{9}$$
$$8 - x = 8 - \frac{40}{9} = \frac{32}{9}$$

It took $\dfrac{40}{9}$ hours to get there and $\dfrac{32}{9}$ hours to return.

29. Let x = number of $10 tickets sold,
$2x$ = number of $8 tickets sold, and
$x + 10$ = number of $12 tickets sold.
$$\$10(x) + \$8(2x) + \$12(x + 10) = \$3122$$
$$10x + 16x + 12x + 120 = 3122$$
$$38x + 120 = 3122$$
$$38x + 120 - 120 = 3122 - 120$$
$$38x = 3002$$
$$\frac{38x}{38} = \frac{3002}{38}$$
$$x = 79$$
$2x = 158$
$x + 10 = 89$
158 tickets at $8 each, 79 tickets at $10 each, and 89 tickets at $12 each were sold.

31.
$$7x + 10 > 80$$
$$7x + 10 - 10 > 80 - 10$$
$$7x > 70$$
$$\frac{7x}{7} > \frac{70}{7}$$
$$x > 10$$

33.
$$4 - 5x < -31$$
$$4 - 4 - 5x < -31 - 4$$
$$-5x < -35$$
$$\frac{-5x}{-5} > \frac{-35}{-5}$$
$$x > 7$$

35.
$$6x - 3 \geq 5(2x + 18)$$
$$6x - 3 \geq 10x + 90$$
$$6x - 10x - 3 \geq 10x - 10x + 90$$
$$-4x - 3 \geq 90$$
$$-4x - 3 + 3 \geq 90 + 3$$
$$-4x \geq 93$$
$$\frac{-4x}{-4} \leq \frac{93}{-4}$$
$$x \leq -\frac{93}{4}$$

37. Let x = number of months of membership.
$$75 + 32.50x \leq 1000$$
$$75 - 75 + 32.50x \leq 1000 - 75$$
$$32.50x \leq 925$$
$$\frac{32.50x}{32.50} \leq \frac{925}{32.50}$$
$$x \leq 28.46153846$$
The person can join for up to 28 months.

39. $\dfrac{82 \text{ miles}}{15 \text{ gallons}}$

41. 2 years = 24 months
$$\frac{4 \text{ months}}{24 \text{ months}} = \frac{1}{6}$$

43.
$$\frac{2}{x} = \frac{14}{63}$$
$$2 \cdot 63 = 14x$$
$$126 = 14x$$
$$\frac{126}{14} = \frac{14x}{14}$$
$$9 = x$$

45.
$$\frac{8}{24} = \frac{24}{x}$$
$$8x = 24 \cdot 24$$
$$8x = 576$$
$$\frac{8x}{8} = \frac{576}{8}$$
$$x = 72$$

47. Let x represent the number of calories burned.

$$\frac{300 \text{ calories}}{12 \text{ minutes}} = \frac{x}{30 \text{ minutes}}$$

$$\frac{300}{12} = \frac{x}{30}$$

$$300 \cdot 30 = 12x$$

$$9000 = 12x$$

$$\frac{9000}{12} = \frac{12x}{12}$$

$$750 = x$$

The person burns 750 calories.

49. The estate was divided into a total of 5 parts.

$$\frac{\$18,000}{5} = \$3600$$

the wife received 3 · $3600 = $10,800.
The son received 2 · $3600 = $7,200.

51. Let y = cost of the deck
  k = constant of variation
  x = area of the deck
  $y = kx$
  $2160 = k(6 \times 9)$
  $2160 = k(54)$
  $\dfrac{2160}{54} = \dfrac{k}{54}$
  $40 = k$
  $y = 40x$
  $y = 40(9 \times 12)$
  $y = 4320$
A 9-ft by 12-ft deck costs $4320 to build.

53. Let y = amperage
  k = constant of variation
  x = resistance
  $y = \dfrac{k}{x}$
  $10 = \dfrac{k}{20}$
  $20 \cdot 10 = 20 \cdot \dfrac{k}{20}$
  $200 = k$
  $y = \dfrac{200}{x}$
  $y = \dfrac{200}{45}$
  $y \approx 4.44$

The amperage is about 4.44 amps.

55. **Step 1** $x^2 - x - 6 = 0$
Step 2 $(x-3)(x+2) = 0$
Step 3 $x - 3 = 0$ $\qquad\qquad$ $x + 2 = 0$
Step 4 $\quad x - 3 = 0$ $\qquad\qquad$ $x + 2 = 0$
$\qquad\quad x - 3 + 3 = 0 + 3$ $\quad$ $x + 2 - 2 = 0 - 2$
$\qquad\qquad\quad x = 3$ $\qquad\qquad\qquad x = -2$

The solutions are -2 and 3.

57. **Step 1** $x^2 - 4x - 21 = 0$
Step 2 $(x-7)(x+3) = 0$
Step 3 $x - 7 = 0$ $\qquad\qquad$ $x + 3 = 0$
Step 4 $\quad x - 7 = 0$ $\qquad\qquad$ $x + 3 = 0$
$\qquad\quad x - 7 + 7 = 0 + 7$ $\quad$ $x + 3 - 3 = 0 - 3$
$\qquad\qquad\quad x = 7$ $\qquad\qquad\qquad x = -3$

The solutions are -3 and 7.

59. **Step 1** $3x^2 + x - 2 = 0$
Step 2 $(3x-2)(x+1) = 0$
Step 3 $3x - 2 = 0$ $\qquad\qquad$ $x + 1 = 0$
Step 4 $\quad 3x - 2 = 0$ $\qquad\qquad$ $x + 1 = 0$
$\qquad\quad 3x - 2 + 2 = 0 + 2$ $\quad$ $x + 1 - 1 = 0 - 1$
$\qquad\qquad\quad 3x = 2$ $\qquad\qquad\qquad x = -1$
$\qquad\qquad\quad \dfrac{3x}{3} = \dfrac{2}{3}$
$\qquad\qquad\qquad x = \dfrac{2}{3}$

The solutions are -1 and $\dfrac{2}{3}$.

61. $x^2 - 5x - 7 = 0$
$a = 1$, $b = -5$, and $c = -7$

$$x = \frac{-b \pm \sqrt{b^2 - 4ac}}{2a}$$

$$= \frac{-(-5) \pm \sqrt{(-5)^2 - 4(1)(-7)}}{2(1)}$$

$$= \frac{5 \pm \sqrt{25 + 28}}{2}$$

$$= \frac{5 \pm \sqrt{53}}{2}$$

The solutions are $\dfrac{5 + \sqrt{53}}{2}$ and $\dfrac{5 - \sqrt{53}}{2}$.

63. $8x^2 + 14x + 4 = 0$

$a = 8$, $b = 14$, and $c = 4$

$x = \dfrac{-b \pm \sqrt{b^2 - 4ac}}{2a}$

$= \dfrac{-14 \pm \sqrt{14^2 - 4(8)(4)}}{2(8)}$

$= \dfrac{-14 \pm \sqrt{196 - 128}}{16}$

$= \dfrac{-14 \pm \sqrt{68}}{16}$

$= \dfrac{-14 \pm \sqrt{4} \cdot \sqrt{17}}{16}$

$= \dfrac{-14 \pm 2\sqrt{17}}{16}$

$= \dfrac{-7 \pm \sqrt{17}}{8}$

The solutions are $\dfrac{-7 + \sqrt{17}}{8}$ and $\dfrac{-7 - \sqrt{17}}{8}$.

65. $4x^2 + 14x - 5 = 0$

$a = 4$, $b = 14$, and $c = -5$

$x = \dfrac{-b \pm \sqrt{b^2 - 4ac}}{2a}$

$= \dfrac{-14 \pm \sqrt{14^2 - 4(4)(-5)}}{2(4)}$

$= \dfrac{-14 \pm \sqrt{196 + 80}}{8}$

$= \dfrac{-14 \pm \sqrt{276}}{8}$

$= \dfrac{-14 \pm \sqrt{4} \cdot \sqrt{69}}{8}$

$= \dfrac{-14 \pm 2\sqrt{69}}{8}$

$= \dfrac{-7 \pm \sqrt{69}}{4}$

The solutions are $\dfrac{-7 + \sqrt{69}}{4}$ and $\dfrac{-7 - \sqrt{69}}{4}$.

67. Let x = first number and $x + 1$ = second number.

$x(x + 1) = 132$

$x^2 + x = 132$

$x^2 + x - 132 = 0$

$(x + 12)(x - 11) = 0$

$x + 12 = 0 \quad\Big|\quad x - 11 = 0$

$x = -12 \quad\Big|\quad x = 11$

When $x = -12$, the numbers are -12 and -11.

When $x = 11$, the numbers are 11 and 12.

Chapter Test

1. $3x - 7y + 2x - 3y + 5$

Combine like terms: $3x + 2x = 5x$

$\qquad\qquad\qquad\qquad -7y - 3y = -10y$

The answer is $5x - 10y + 5$.

3. $3x^2 - 2x + 6 = 3(-5)^2 - 2(-5) + 6$

$\qquad\qquad\quad = 3(25) - 2(-5) + 6$

$\qquad\qquad\quad = 75 + 10 + 6$

$\qquad\qquad\quad = 91$

5. $3x - 5(2x + 10) = -59$

$3x - 10x - 50 = -59$

$-7x - 50 = -59$

$-7x - 50 + 50 = -59 + 50$

$-7x = -9$

$\dfrac{-7x}{-7} = \dfrac{-9}{-7}$

$x = \dfrac{9}{7}$

$\left\{ \dfrac{9}{7} \right\}$

7. $F = \dfrac{mv^2}{r}$

$Fr = r \cdot \dfrac{mv^2}{r}$

$Fr = mv^2$

$\dfrac{Fr}{F} = \dfrac{mv^2}{F}$

$r = \dfrac{mv^2}{F}$

9. $\qquad 4 - 3x \geq x + 10$

$4 - 3x - x \geq x - x + 10$

$4 - 4x \geq 10$

$4 - 4 - 4x \geq 10 - 4$

$-4x \geq 6$

$\dfrac{-4x}{-4} \leq \dfrac{6}{-4}$

$x \leq -\dfrac{3}{2}$

$\left\{ x \,\Big|\, x \leq -\dfrac{3}{2} \right\}$

11. $\dfrac{x}{9} = \dfrac{16}{36}$

$36x = 9 \cdot 16$

$36x = 144$

$\dfrac{36x}{36} = \dfrac{144}{36}$

$x = 4$

13. $(x-8)(2x+3) = x \cdot 2x + x \cdot 3 + (-8)(2x) + (-8)(3)$

$$= 2x^2 + 3x - 16x - 24$$

$$= 2x^2 - 13x - 24$$

15. **Step 1** $x^2 - 14x - 51 = 0$

Step 2 $(x+3)(x-17) = 0$

Step 3 $x + 3 = 0 \qquad x - 17 = 0$

Step 4 $x + 3 = 0 \qquad x - 17 = 0$

$$x = -3 \qquad\qquad x = 17$$

The solution set is $\{-3, 17\}$.

17. **Step 1** $6x^2 + x - 12 = 0$

Step 2 $(2x+3)(3x-4) = 0$

Step 3 $2x + 3 = 0 \qquad 3x - 4 = 0$

Step 4 $\quad 2x + 3 = 0 \qquad\qquad 3x - 4 = 0$

$$2x + 3 - 3 = 0 - 3 \qquad 3x - 4 + 4 = 0 + 4$$

$$2x = -3 \qquad\qquad 3x = 4$$

$$\frac{2x}{2} = \frac{-3}{2} \qquad\qquad \frac{3x}{4} = \frac{4}{3}$$

$$x = -\frac{3}{2} \qquad\qquad x = \frac{4}{3}$$

The solution set is $\left\{-\dfrac{3}{2}, \dfrac{4}{3}\right\}$.

19. $5x^2 + 2x - 3 = 0$

$a = 5$, $b = 2$, and $c = -3$

$$x = \frac{-b \pm \sqrt{b^2 - 4ac}}{2a}$$

$$= \frac{-2 \pm \sqrt{2^2 - 4(5)(-3)}}{2(5)}$$

$$= \frac{-2 \pm \sqrt{4 + 60}}{10}$$

$$= \frac{-2 \pm \sqrt{64}}{10}$$

$$= \frac{-2 \pm 8}{10}$$

$$x = \frac{-2 + 8}{10} \quad \text{or} \quad x = \frac{-2 - 8}{10}$$

$$= \frac{6}{10} \qquad\qquad = \frac{-10}{10}$$

$$= \frac{3}{5} \qquad\qquad = -1$$

$$\left\{-1, \frac{3}{5}\right\}$$

21. Let x = amount invested at 6% and

$3000 - x$ = amount invested at 8%.

$$0.06x + 0.08(3000 - x) = 190$$

$$0.06x + 240 - 0.08x = 190$$

$$-0.02x + 240 = 190$$

$$-0.02x + 240 - 40 = 190 - 240$$

$$-0.02x = -50$$

$$\frac{-0.02x}{-0.02} = \frac{-50}{-0.02}$$

$$x = 2500$$

$3000 - x = 500$

$2500 is invested at 6% and $500 is invested at 8%.

23. Let x = number of minutes to bike 210 miles.

$$\frac{2 \text{ miles}}{12.5 \text{ minutes}} = \frac{210 \text{ miles}}{x}$$

$$2x = 12.5(210)$$

$$2x = 2625$$

$$\frac{2x}{2} = \frac{2625}{2}$$

$$x = 1312.5$$

$$\frac{1312.5 \text{ minutes}}{60 \text{ minutes}} = 21.875 \text{ or about 21.9 hours}$$

25. Let y = number of days to do the job

k = constant of variation

x = number of people working

$$y = \frac{k}{x}$$

$8 \text{ hours} = \dfrac{1}{3} \text{ day}$

$$\frac{1}{3} = \frac{k}{12}$$

$$12 \cdot \frac{1}{3} = 12 \cdot \frac{k}{12}$$

$$4 = k$$

$$y = \frac{4}{x}$$

$$y = \frac{4}{8}$$

$$y = \frac{1}{2}$$

It takes $\dfrac{1}{2}$ day or 12 hours to complete the job.

27. Let x = width of the board and
$x + 4$ = length of the board.

$$\text{Area} = \text{length} \times \text{width}$$
$$96 = x(x+4)$$
$$96 = x^2 + 4x$$
$$x^2 + 4x - 96 = 0$$
$$(x+12)(x-8) = 0$$

$$x + 12 = 0 \qquad x - 8 = 0$$
$$x = -12 \qquad x = 8$$

The width is 8 inches and the length is 12 inches.

8 Additional Topics in Algebra

Exercise Set 8-1

1. (−2, −5) is plotted by starting at the origin and moving two units left and five units down.

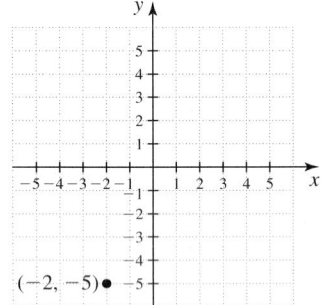

3. (−6, 4) is plotted by starting at the origin and moving six units left and four units up.

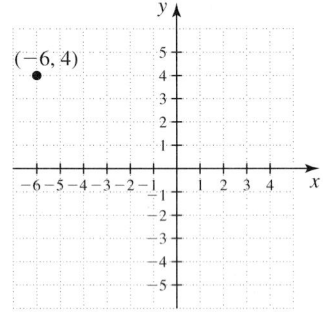

5. (6, 0) is plotted by starting at the origin and moving six units right.

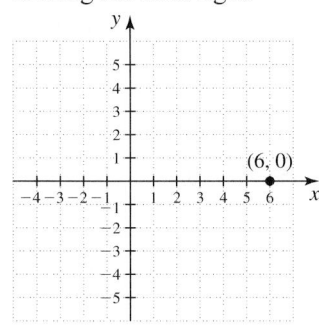

7. (0, 3) is plotted by starting at the origin and moving three units up.

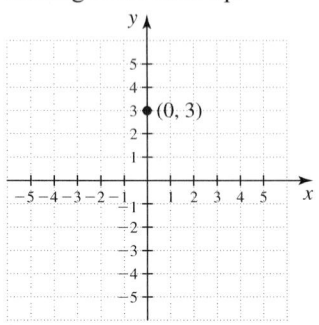

9. (5.6, −3.2) is plotted by starting at the origin and moving 5.6 units right and 3.2 units down.

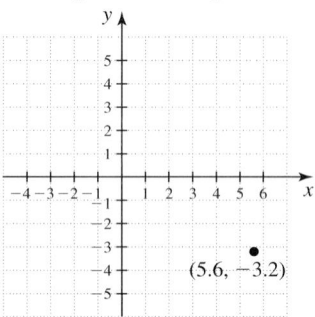

11. (−6, −10) is plotted by starting at the origin and moving six units left and ten units down.

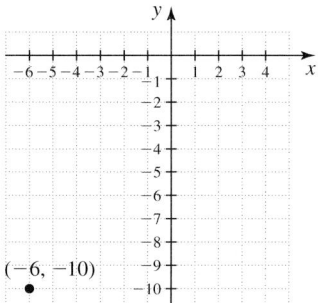

13. (0, 0) is plotted at the origin.

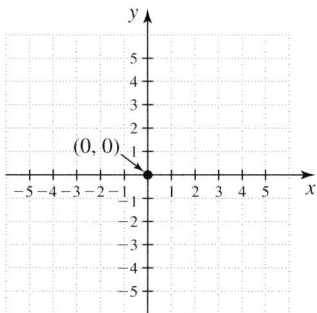

15. Substitute $x = 2$, $x = 3$, and $x = 4$ in the equation $5x + y = 20$ and solve each equation for y. The three points are (2, 10), (3, 5), and (4, 0). Plot the points on the plane and draw a straight line through them.

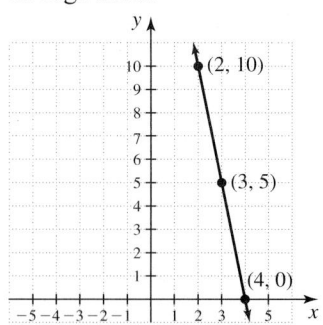

17. Substitute $x = 4$, $x = 5$, and $x = 6$ in the equation $3x - y = 15$ and solve each equation for y. The three points are $(4, -3)$, $(5, 0)$, and $(6, 3)$. Plot the points on the plane and draw a straight line through them.

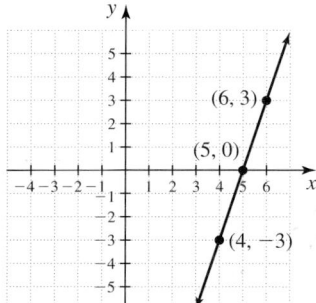

19. Substitute $x = -2$, $x = 0$, and $x = 7$ in the equation $4x + 7y = 28$ and solve each equation for y. The three points are $\left(-2, 5\frac{1}{7}\right)$, $(0, 4)$, and $(7, 0)$.

Plot the points on the plane and draw a straight line through them.

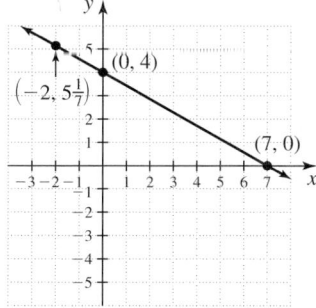

21. Substitute $x = -6$, $x = -2$, and $x = 1$ in the equation $2x + 7y = -12$ and solve each equation for y. The three points are $(-6, 0)$, $\left(-2, -\frac{8}{7}\right)$, and $(1, -2)$. Plot the points on the plane and draw a straight line through them.

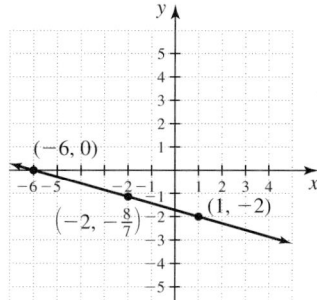

23. Substitute $x = 0$, $x = 4$, and $x = 6$ in the equation $6x - 4y = 28$ and solve each equation for y. The three points are $(0, -7)$, $(4, -1)$, and $(6, 2)$. Plot the points on the plane and draw a straight line through them

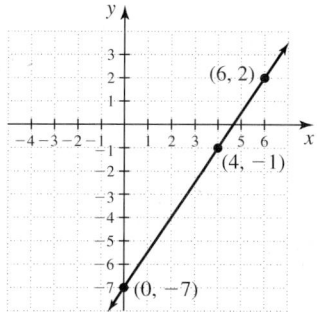

25. $m = \dfrac{y_2 - y_1}{x_2 - x_1} = \dfrac{7 - (-2)}{6 - (-3)} = \dfrac{9}{9} = 1$

The slope is 1.

27. $m = \dfrac{y_2 - y_1}{x_2 - x_1} = \dfrac{9 - 10}{4 - 2} = \dfrac{-1}{2} = -0.5$

The slope is -0.5.

29. $m = \dfrac{y_2 - y_1}{x_2 - x_1} = \dfrac{-2 - (-5)}{-9 - (-4)} = \dfrac{3}{-5} = -\dfrac{3}{5}$

The slope is $-\dfrac{3}{5}$.

31. $m = \dfrac{y_2 - y_1}{x_2 - x_1} = \dfrac{-2.8 - (-6.1)}{3.4 - 2} = \dfrac{3.3}{1.4}$

The slope is about 2.4.

33. To find the x intercept, let $y = 0$ and solve for x.
$$3x + 4y = 24$$
$$3x + 4(0) = 24$$
$$3x = 24$$
$$x = 8$$
To find the y intercept, let $x = 0$ and solve for y.
$$3x + 4y = 24$$
$$3(0) + 4y = 24$$
$$4y = 24$$
$$y = 6$$
Hence, the x intercept $= (8, 0)$ and the y intercept $= (0, 6)$.

35. To find the x intercept, let $y = 0$ and solve for x.
$$-5x - 6y = 30$$
$$-5x - 6(0) = 30$$
$$-5x = 30$$
$$x = -6$$
To find the y intercept, let $x = 0$ and solve for y.
$$-5x - 6y = 30$$
$$-5(0) - 6y = 30$$
$$-6y = 30$$
$$y = -5$$

Hence, the x intercept $= (-6, 0)$ and the y intercept $= (0, -5)$.

37. To find the x intercept, let $y = 0$ and solve for x.
$$2x - y = 18$$
$$2x - 0 = 18$$
$$2x = 18$$
$$x = 9$$
To find the y intercept, let $x = 0$ and solve for y.
$$2x - y = 18$$
$$2(0) - y = 18$$
$$-y = 18$$
$$y = -18$$
Hence, the x intercept = (9, 0) and the
y intercept = (0, −18).

39. To find the x intercept, let $y = 0$ and solve for x.
$$5x - 2y = 15$$
$$5x - 2(0) = 15$$
$$5x = 15$$
$$x = 3$$
To find the y intercept, let $x = 0$ and solve for y.
$$5x - 2y = 15$$
$$5(0) - 2y = 15$$
$$-2y = 15$$
$$y = -7.5$$
Hence, the x intercept = (3, 0) and the
y intercept = (0, −7.5).

41. Solve the equation for y.
$$7x + 5y = 35$$
$$5y = -7x + 35$$
$$\frac{5y}{5} = \frac{-7x}{5} + \frac{35}{5}$$
$$y = -\frac{7}{5}x + 7$$

The slope $= -\dfrac{7}{5}$ and the y intercept = (0, 7). To
graph, use (0, 7) as one point and from this point
move down 7 units and right 5 units. The second
point is (5, 0). Draw a line through the two
points.

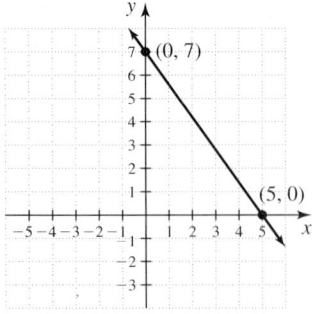

43. Solve the equation for y.
$$x - 4y = 16$$
$$-4y = -x + 16$$
$$\frac{-4y}{-4} = \frac{-x}{-4} + \frac{16}{-4}$$
$$y = \frac{1}{4}x - 4$$

The slope $= \dfrac{1}{4}$ and the y intercept = (0, −4). To
graph, use (0, −4) as one point. From this point
move up 1 unit (or 2 units) and right 4 units (or
8 units). A second point is (8, −2). Draw a line
through the two points.

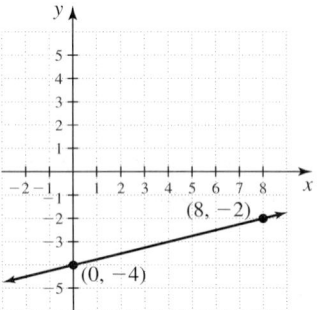

45. Solve the equation for y.
$$8x - 3y = 24$$
$$-3y = -8x + 24$$
$$\frac{-3y}{-3} = \frac{-8x}{-3} + \frac{24}{-3}$$
$$y = \frac{8}{3}x - 8$$

The slope $= \dfrac{8}{3}$ and the y intercept = (0, −8). To
graph, use (0, −8) as one point. From this point
move up 8 units and right 3 units. A second
point is (3, 0). Draw a line through the two
points.

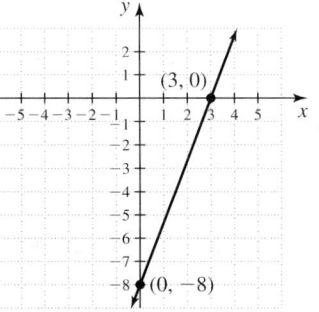

47. Solve the equation for y.
$$2x - y = 19$$
$$-y = -2x + 19$$
$$y = 2x - 19$$

The slope = 2 and the y intercept = (0, −19).
To graph, use (0, −19) as one point. From this
point move up 2 units and right 1 unit. Draw a
line through the two points. The graph shows
two other points on the line.

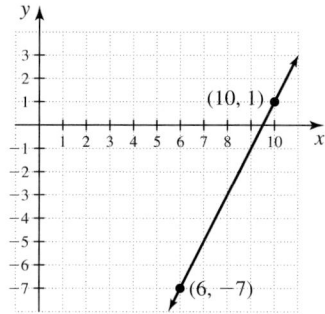

49. The graph of $x = -3$ is a vertical line passing through $(-3, 0)$ on the x axis.

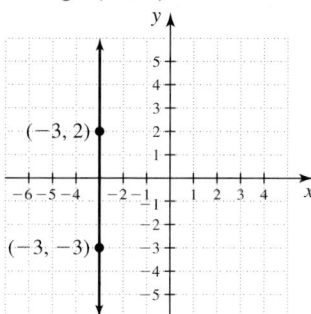

51. The graph of $y = 6$ is a horizontal line passing through $(0, 6)$ on the y axis.

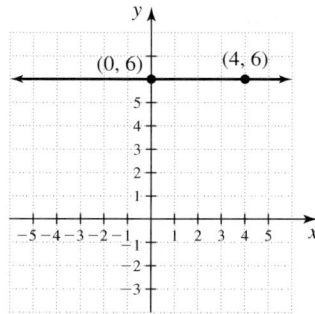

53. Let y = cost of running the ad and let x = number of weeks. Since it costs \$6.50 per week, the cost of x weeks is $6.5x$. This together with the setup charge of \$50.00 gives us the equation $y = 6.5x + 50$.

  (a) For 3 weeks,
     $y = 6.5(3) + 50$
     $y = \$69.50$

  (b) For 5 weeks,
     $y = 6.5(5) + 50$
     $y = \$82.50$

  (c) For 10 weeks,
     $y = 6.5(10) + 50$
     $y = \$115$

55. Let y = rental cost and let x = number of miles. Since it costs \$1.10 per mile, the cost of x miles is $1.1x$. This, together with the daily charge of \$40 for one day, gives us the equation $y = 1.1x + 40$.

  (a) For 63 miles,
     $y = 1.1(63) + 40$
     $y = \$109.30$

  (b) For 42 miles,
     $y = 1.1(42) + 40$
     $y = \$86.20$

  (c) For 127 miles,
     $y = 1.1(127) + 40$
     $y = \$179.70$

57. Substitute 3 for x in the equation.
    $y = 10.5x$
    $y = 10.5(3)$
    $y = 31.5$
There will be 31.5 thousand or 31,500 civilians in the military during the year that is 3 years from now.

59. Substitute 5 for x in the equation.
    $y = 0.5x + 35.3$
    $y = 0.5(5) + 35.3$
    $y = 37.8$
There will be 37.8 million people over 65 living 5 years from now.

61. Quadrants I, II, III, and IV, respectively, have points whose coordinates are $(+, +)$, $(-, +)$, $(-, -)$, $(+, -)$.

63. Let $(x_0,\ 0)$ = the x intercept. To find x_0, let $y = 0$ in the equation of the line and solve for x. Let $(0,\ y_0)$ = the y intercept. To find y_0, let $x = 0$ in the equation of the line and solve for y.

65. If the equation reduces to the form y = constant, then the line is horizontal.

67. The denominator of the fraction for the slope will always be zero.

69. Let (x_1, y_1) and (x_2, y_2) be two points on a line. Then $y_2 - y_1$ is the vertical distance between the points and $x_2 - x_1$ is the horizontal distance between the points. By the Pythagorean theorem, $c^2 = a^2 + b^2$, the distance, d, between the two points on the line is
$$d^2 = (y_2 - y_1)^2 + (x_2 - x_1)^2 \text{ or}$$
$$d = \sqrt{(y_2 - y_1)^2 + (x_2 - x_1)^2}.$$

Exercise Set 8-2

1. Draw the graphs for both equations on the Cartesian plane.

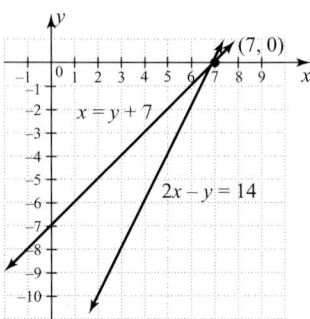

The point of intersection is (7, 0). Hence, the solution is {(7, 0)}.

3. Draw the graphs for both equations on the Cartesian plane.

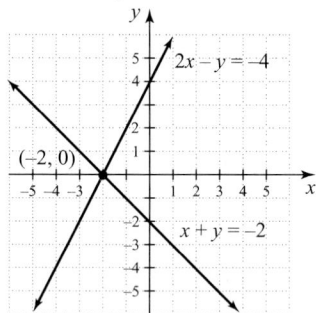

The point of intersection is (−2, 0). Hence, the solution is {(−2, 0)}.

5. Draw the graphs for both equations on the Cartesian plane.

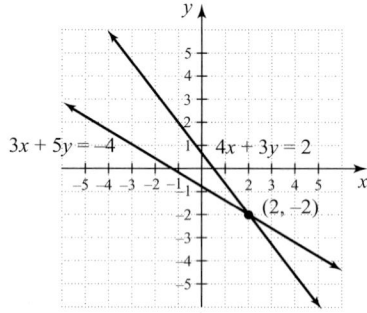

The point of intersection is (2, −2). Hence, the solution is {(2, −2)}.

7. Draw the graphs for both equations on the Cartesian plane.

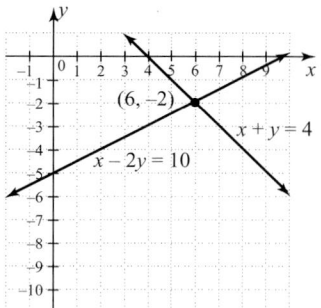

The point of intersection is (6, −2). Hence the solution is {(6, −2)}.

9. Select one equation and solve for one variable.

$x + y = 0$
$x = -y$

Substitute $x = -y$ in the second equation and solve for y.

$x - 2y + 9 = 0$
$-y - 2y + 9 = 0$
$-3y = -9$
$\dfrac{-3y}{-3} = \dfrac{-9}{-3}$
$y = 3$

Substitute $y = 3$ in one equation and solve for x.

$x + y = 0$
$x + 3 = 0$
$x = -3$

The solution set is {(−3, 3)}.

11. Select one equation and solve for one variable.

$x - 3y = 7$
$x = 3y + 7$

Substitute $x = 3y + 7$ in the second equation and solve for y.

$4x + 3y = 13$
$4(3y + 7) + 3y = 13$
$12y + 28 + 3y = 13$
$15y = -15$
$\dfrac{15y}{15} = \dfrac{-15}{15}$
$y = -1$

Substitute $y = -1$ in one equation and solve for x.

$x - 3y = 7$
$x - 3(-1) = 7$
$x + 3 = 7$
$x = 4$

The solution set is {(4, −1)}.

13. Select one equation and solve for one variable.

$4x + 3y = 24$
$3y = 24 - 4x$
$y = 8 - \dfrac{4x}{3}$

Substitute $y = 8 - \dfrac{4x}{3}$ in the second equation and solve for x.

$$3x + 5y = 22$$
$$3x + 5\left(8 - \frac{4x}{3}\right) = 22$$
$$3x + 40 - \frac{20x}{3} = 22$$
$$3x - \frac{20x}{3} = -18$$
$$3 \cdot 3x - \frac{3}{1} \cdot \frac{20x}{3} = 3(-18)$$
$$9x - 20x = -54$$
$$-11x = -54$$
$$x = \frac{54}{11}$$

Substitute $x = \dfrac{54}{11}$ in one equation and solve for y.

$$4x + 3y = 24$$
$$4\left(\frac{54}{11}\right) + 3y = 24$$
$$\frac{216}{11} + 3y = 24$$
$$3y = \frac{264}{11} - \frac{216}{11}$$
$$3y = \frac{48}{11}$$
$$y = \frac{16}{11}$$

The solution set is $\left\{\left(\dfrac{54}{11}, \dfrac{16}{11}\right)\right\}$.

15. Substitute $x = y + 2$ in the first equation and solve for y.
$$2x - 3y = 1$$
$$2(y + 2) - 3y = 1$$
$$2y + 4 - 3y = 1$$
$$-y = -3$$
$$y = 3$$
Substitute $y = 3$ in one equation and solve for x.
$$x = y + 2$$
$$x = 3 + 2$$
$$x = 5$$
The solution set is $\{(5, 3)\}$.

17. Write both equations in the form $ax + by = c$.
$$3x - 2y = 10$$
$$-3x + 9y = -7$$

Add the equations to eliminate the x variable. Then solve for y.
$$3x - 2y = 10$$
$$\underline{-3x + 9y = -7}$$
$$7y = 3$$
$$y = \frac{3}{7}$$

Substitute $y = \dfrac{3}{7}$ in one equation and solve for x.

$$3x = 2y + 10$$
$$3x = 2\left(\frac{3}{7}\right) + 10$$
$$3x = \frac{6}{7} + \frac{70}{7}$$
$$3x = \frac{76}{7}$$
$$\frac{1}{3} \cdot 3x = \frac{1}{3} \cdot \frac{76}{7}$$
$$x = \frac{76}{21}$$

The solution set is $\left\{\left(\dfrac{76}{21}, \dfrac{3}{7}\right)\right\}$.

19. Write both equations in the form $ax + by = c$.
$$3x + 4y = 2$$
$$-x + y = -3$$
Multiply the second equation by 3. Add the equations to eliminate the x variable and solve for y.
$$3x + 4y = 2$$
$$\underline{-3x + 3y = -9}$$
$$7y = -7$$
$$y = -1$$
Substitute $y = -1$ in one equation and solve for x.
$$y = x - 3$$
$$-1 = x - 3$$
$$2 = x$$
The solution set is $\{(2, -1)\}$.

21. Multiply the first equation by -3. Then add the equations.
$$-15x + 6y = -33$$
$$\underline{15x - 6y = 33}$$
$$0 = 0$$

Since the variables are eliminated and $0 = 0$ is true, the system is dependent. The solution set is $\{(x, y) \mid 5x - 2y = 11\}$.

23. Write both equations in the form $ax + by = c$.
$$x - 3y = -2$$
$$5x - 4y = 12$$

Multiply the first equation by -5. Add the equations to eliminate the x variable. Then solve for y.
$$-5x + 15y = 10$$
$$\underline{5x - 4y = 12}$$
$$11y = 22$$
$$y = 2$$
Substitute $y = 2$ in one equation and solve for x.
$$x = 3y - 2$$
$$x = 3(2) - 2$$
$$x = 6 - 2$$
$$x = 4$$
The solution set is $\{(4, 2)\}$.

25. Select one equation and solve for one variable.
$$4x + y = 2$$
$$y = 2 - 4x$$

Substitute $y = 2 - 4x$ in the second equation and solve for x.
$$7x + 3y = 1$$
$$7x + 3(2 - 4x) = 1$$
$$7x + 6 - 12x = 1$$
$$-5x = -5$$
$$x = 1$$

Substitute $x = 1$ in one equation and solve for y.
$$4x + y = 2$$
$$4(1) + y = 2$$
$$4 + y = 2$$
$$y = -2$$

The system is consistent. The solution set is $\{(1, -2)\}$.

27. Select one equation and solve for one variable.
$$x - 6y = 19$$
$$x = 6y + 19$$

Substitute $x = 6y + 19$ in the second equation and solve for y.
$$2x + 7y = 0$$
$$2(6y + 19) + 7y = 0$$
$$12y + 38 + 7y = 0$$
$$19y = -38$$
$$y = -2$$

Substitute $y = -2$ in one equation and solve for x.
$$x - 6y = 19$$
$$x - 6(-2) = 19$$
$$x + 12 = 19$$
$$x = 7$$

The system is consistent. The solution set is $\{(7, -2)\}$.

29. Multiply the first equation by -5 and the second equation by 3. Add the equations to eliminate the x variable. Then solve for y.
$$-15x + 25y = -25$$
$$\underline{15x - 21y = 3}$$
$$4y = -22$$
$$y = -\frac{22}{4}$$
$$y = -\frac{11}{2}$$

Substitute $y = -\frac{11}{2}$ in one equation and solve for x.
$$3x - 5y = 5$$
$$3x - 5\left(-\frac{11}{2}\right) = 5$$
$$3x + \frac{55}{2} = 5$$
$$3x = -\frac{45}{2}$$
$$x = -\frac{15}{2}$$

The system is consistent. The solution set is
$$\left\{\left(-\frac{15}{2}, -\frac{11}{2}\right)\right\}.$$

31. Select one equation and solve for one variable.
$$4x - y = 11$$
$$-y = -4x + 11$$
$$y = 4x - 11$$

Substitute $y = 4x - 11$ in the second equation and solve for x.
$$7x + 3y = 8$$
$$7x + 3(4x - 11) = 8$$
$$7x + 12x - 33 = 8$$
$$19x = 41$$
$$x = \frac{41}{19}$$

Substitute $x = \frac{41}{19}$ in one equation and solve for y.
$$4x - y = 11$$
$$4\left(\frac{41}{19}\right) - y = 11$$
$$\frac{164}{19} - y = 11$$
$$-y = \frac{45}{19}$$
$$y = -\frac{45}{19}$$

The system is consistent. The solution set is
$$\left\{\left(\frac{41}{19}, -\frac{45}{19}\right)\right\}.$$

33. Let x = the older sister's age and let y = the younger sister's age.
The system is
$$x + y = 32$$
$$4x - 2y = 38$$

Solve by the addition method.
$$2x + 2y = 64$$
$$\underline{4x - 2y = 38}$$
$$6x = 102$$
$$x = 17$$

Substitute $x = 17$ in one equation and solve for y.
$$x + y = 332$$
$$17 + y = 32$$
$$y = 15$$

The ages are 15 and 17.

35. Let x = number of adults and let y = number of children.
The system is
$$x + y = 40$$
$$12x + 8y = 420$$

Solve by the substitution method.
$$x + y = 40$$
$$y = 40 - x$$

Substitute $y = 40 - x$ in the second equation.

$$12x + 8y = 420$$
$$12x + 8(40 - x) = 420$$
$$12x + 320 - 8x = 420$$
$$4x = 100$$
$$x = 25$$

Substitute $x = 25$ in one equation and solve for y.
$$x + y = 40$$
$$25 + y = 40$$
$$y = 15$$

There were 25 adults and 15 children who rode the train.

37. Let x = age of older student and let y = age of younger student.
The system is
$$x - y = 4$$
$$x + y = 42$$

Solve by the addition method.
$$\begin{aligned} x - y &= 4 \\ x + y &= 42 \\ \hline 2x &= 46 \end{aligned}$$
$$x = 23$$

Substitute $x = 23$ in one equation and solve for y.
$$x + y = 42$$
$$23 + y = 42$$
$$y = 19$$

The ages of the students are 19 years and 23 years.

39. Let x = cost of sandwich and let y = cost of French fries.
The system is
$$3x + 2y = 8.87$$
$$5x + 4y = 15.55$$

Multiply the first equation by -2 and solve by the addition method.
$$\begin{aligned} -6x - 4y &= -17.74 \\ 5x + 4y &= 15.55 \\ \hline -x &= -2.19 \end{aligned}$$
$$x = 2.19$$

Substitute $x = 2.19$ in one equation and solve for y.
$$3x + 2y = 8.87$$
$$3(2.19) + 2y = 8.87$$
$$6.57 + 2y = 8.87$$
$$2y = 2.30$$
$$y = 1.15$$

The cost for one sandwich is $2.19 and one order of French fries is $1.15.

41. Let x = number of student tickets and let y = number of general admission tickets.
The system is
$$x + y = 500$$
$$5x + 8y = 3034$$

Multiply the first equation by -5 and solve by the addition method.

$$\begin{aligned} -5x - 5y &= -2500 \\ 5x + 8y &= 3034 \\ \hline 3y &= 534 \\ y &= 178 \end{aligned}$$

Substitute $y = 178$ in one equation and solve for x.
$$x + y = 500$$
$$x + 178 = 500$$
$$x = 322$$

There were 322 students attending the concert and 178 general admission tickets sold.

43. Algebraically: The system has exactly one solution, or else an infinite number of solutions. Geometrically: The lines intersect as exactly one point, or else the lines coincide.

45. Algebraically: Any solution for either line is also a solution for the other; hence, an infinite number of solutions exist. Geometrically: The two lines coincide; that is, they are the same line.

47. Say the two variables are y and z. In either equation solve for one variable, say y, and substitute into the other equation to solve for z. Substitute this actual value of z into either original equation, to solve for the (actual, numerical value of y.

49. System is inconsistent if the solution procedure leads to some constant = some different constant (e.g., $20 = 14$).

51. Answers will vary.

Exercise Set 8-3

1. Graph the solid (since the inequality sign is $\geq$) line $2x - y = 5$ using two points, such as $(3, 1)$ and $(0, -5)$. Select $(0, 0)$ as a test point.
$$2x - y \geq 5$$
$$2(0) - 0 \geq 5?$$
$$0 \geq 5 \text{ False}$$
Since this is a false statement, shade the area that does not contain the test point.

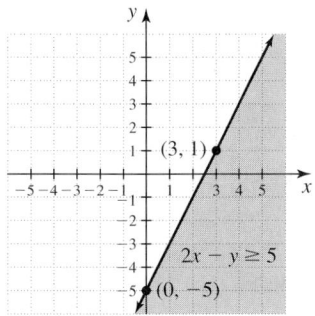

3. Graph the solid (since the inequality sign is ≤)
line $3x - 4y = 12$ using two points, such as
$(0, -3)$ and $(4, 0)$. Select $(0, 0)$ as a test point.
$$3x - 4y \le 12$$
$$3(0) - 4(0) \le 12?$$
$$0 \le 12 \text{ True}$$
Since this is a true statement, shade the area that
contains the test point.

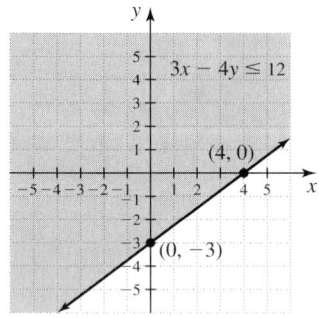

5. To graph $x \ge 5$, use $x = 5$ and draw a solid
vertical line passing through 5 on the x axis.
Then use $(0, 0)$ as a test point. Since $0 \ge 5$ is a
false statement, shade the half plane that does
not contain $(0, 0)$.

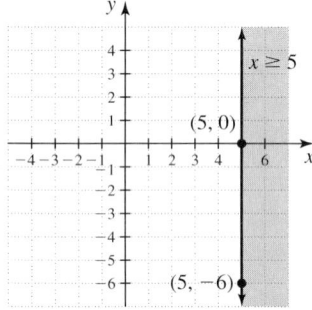

7. To graph $y \le 0$, use $y = 0$ and draw a solid
horizontal line passing through 0 on the y axis.
Then use $(0, -1)$ as a test point. Since $-1 \le 0$ is a
true statement, shade the half plane below the
line.

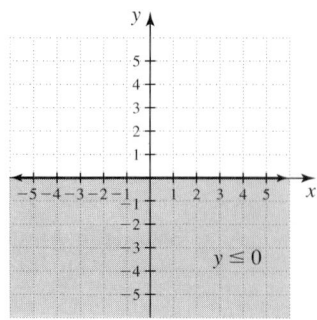

9. Graph the dashed (since the inequality sign is <)
line $3x + y = -6$ using two points, such as $(-2, 0)$
and $(0, -6)$. Select $(0, 0)$ as a test point.
$$3x + y < -6$$
$$3(0) + 0 < -6?$$
$$0 < -6 \text{ False}$$
Since this is a false statement, shade the area that

does not contain the test point.

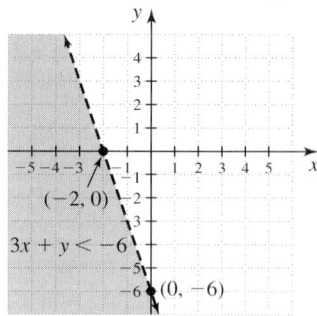

11. Graph $2x - y \le 6$ using a solid line (since the
inequality sign is ≤) and two points, such as
$(0, -6)$ and $(3, 0)$. Select $(0, 0)$ as a test point.
$$2x - y \le 6$$
$$2(0) - 0 \le 6?$$
$$0 \le 6 \text{ True}$$
Since this is a true statement, shade the area that
contains the test point.
Graph $x + y > 3$ using a dashed line (since the
inequality sign is >) and two points, such as
$(3, 0)$ and $(0, 3)$. Select $(0, 0)$ as a test point.
$$x + y > 3$$
$$0 + 0 > 3?$$
$$0 > 3 \text{ False}$$
Since this is a false statement, shade the area that
does not contain the test point. The solution is
the intersection of the two half planes and part of
the line $2x - y = 6$.

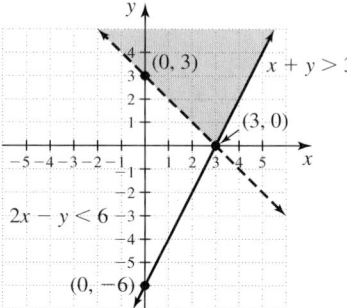

13. To graph $x \ge -2$, use $x = -2$ and draw a solid
vertical line passing through -2 on the x axis.
Then use $(0, 0)$ as a test point. Since $0 \ge -2$ is a
true statement, shade the half plane to the right
of the line.
To graph $y < -3$, use $y = -3$ and draw a dashed
horizontal line passing through -3 on the y axis.
Then use $(0, 0)$ as a test point. Since $0 < -3$ is a
false statement, shade the half plane below the
line.
The solution is the intersection of the two half
planes and part of the line $x = -2$.

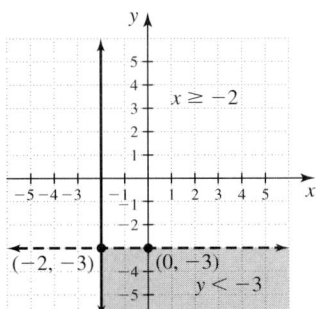

15. Graph $5x - 3y \geq 15$ using a solid line (since the inequality sign is $\geq$) and two points, such as $(0, -5)$ and $(3, 1)$. Select $(0, 0)$ as a test point.
$$5x - 3y \geq 15$$
$$5(0) - 3(0) \geq 15?$$
$$0 \geq 15 \text{ False}$$
Since this is a false statement, shade the area that does not contain the test point.
Graph $3x - 8y \geq 12$ using a solid line (since the inequality symbol is $\geq$) and two points, such as $(0, -1.5)$ and $(4, 0)$. Select $(0, 0)$ as a test point.
$$3x - 8y \geq 12$$
$$3(0) - 8(0) \geq 12?$$
$$0 \geq 12 \text{ False}$$
Since this is a false statement, shade the area that does not contain the test point.
The solution is the intersection of the two half planes and parts of the lines $5x - 3y = 15$ and $3x - 8y = 12$.

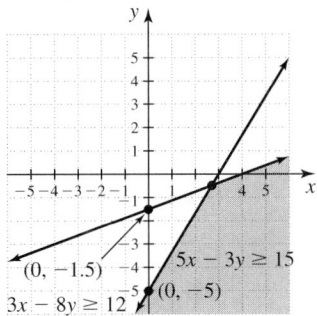

17. Graph $x + y \geq 7$ using a solid line (since the inequality sign is $\geq$) and two points, such as $(0, 7)$ and $(1, 6)$. Select $(0, 0)$ as a test point.
$$x + y \geq 7$$
$$0 + 0 \geq 7?$$
$$0 \geq 7 \text{ False}$$
Since this is a false statement, shade the half plane that does not contain the test point.
Graph $x - y < -5$ using a dashed line (since the inequality sign is $<$) and two points, such as $(5, 10)$ and $(1, 6)$. Select $(0, 0)$ as a test point.
$$x - y < -5$$
$$0 - 0 < -5?$$
$$0 < -5 \text{ False}$$
Since this is a false statement, shade the half plane that does not contain the test point.
The solution is the intersection of the two half planes and part of the line $x + y = 7$.

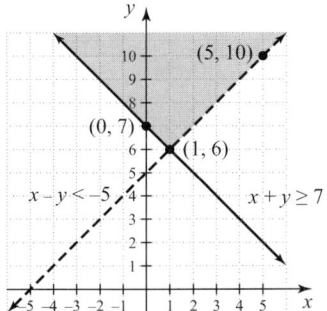

19. Graph $4x - 2y < 6$ using a dashed line (since the inequality sign is $<$) and two points, such as $(2, 1)$ and $(0, -3)$. Select $(0, 0)$ as a test point.
$$4x - 2y < 6$$
$$4(0) - 2(0) < 6?$$
$$0 < 6 \text{ True}$$
Since this is a true statement, shade the area above the line.
Graph $8x - 4y > -3$ using a dashed line (since the inequality sign is $>$) and two points, such as $\left(0, \dfrac{3}{4}\right)$ and $\left(-2, -\dfrac{13}{4}\right)$. Select $(0, 0)$ as a test point.
$$8x - 4y > -3$$
$$8(0) - 4(0) > -3?$$
$$0 > -3 \text{ True}$$
Since this is a true statement, shade the area below the line.
The solution is the intersection of the two half planes.

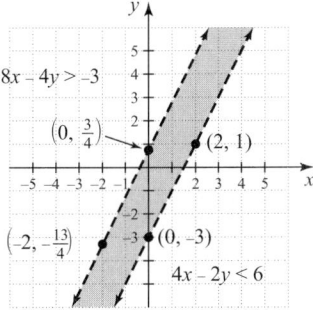

21. A solid line indicates all points of it are part of (belong to) the graph. A dashed line indicates no points of it are part of (belong to) the graph.

23. When the signs of both inequalities are a combination of $\geq$ or $\leq$.

25. Graph $x + y > 6$ using a dashed line (since the inequality sign is $>$) and two points such as $(6, 0)$ and $(0, 6)$. Select $(0, 0)$ as a test point.
$$x + y > 6$$
$$0 + 0 > 6?$$
$$0 > 6 \text{ False}$$
Since this is a false statement, shade the area above the line.
Graph $x + y < -3$ using a dashed line (since the inequality sign is $<$) and two points, such as

(−3, 0) and (0, −3). Select (0, 0) as a test point.

$x + y < -3$

$0 + 0 < -3?$

   $0 < -3$ False

Since this is a false statement, shade the area below the line.

The two half planes do not intersect. The solution is ∅.

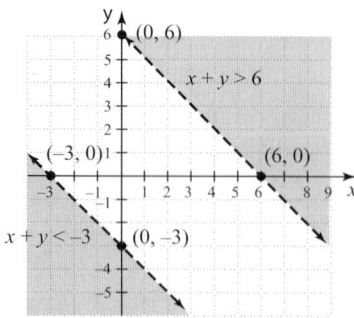

Exercise Set 8-4

1. Since at most 50 calculators can be manufactured in one day, the constraint is

$x \leq 50$

where x = number of calculators.

3. Since the company wants to manufacture twice as many pens as pencils, the constraint is

$x = 2y$

where x = number of pens and y = number of pencils.

5. Since the profit on a desktop computer is $85.00 and the profit on a laptop computer is $130.00, the profit function is

$P = 85d + 130l$

where d = number of desktop computers sold and

l = number of laptop computers sold.

7. Since the profit on each turkey is $4.00 and the profit on each ham is $3.00, the profit function is

$P = 4t + 3h$

where t = number of turkeys sold and h = number of hams sold.

9. Since it takes 2 hours to assemble a VCR and 3 hours to assemble a CD player, and each employee can work at most 6 hours, the constraint is

$2v + 3c \leq 6$

where v = number of VCRs assembled and c = number of CD players assembled.

11. $P = 30x + 40y$

(0, 0) $P = 30(0) + 40(0) = 0 + 0 = \$0$

(5, 6) $P = 30(5) + 40(6) = 150 + 240 = \$390$

(10, 0) $P = 30(10) + 40(0) = 300 + 0 = \$300$

(0, 8) $P = 30(0) + 40(9) = 0 + 360 = \$360$

Hence, vertex (5, 6) gives a maximum profit of $390.

13. $P = 15x + 5y$

(0, 0) $P = 15(0) + 5(0) = 0 + 0 = \$0$

(4, 8) $P = 15(4) + 5(8) = 60 + 40 = \$100$

(0, 6) $P = 15(0) + 5(6) = 0 + 30 = \$30$

(7, 0) $P = 15(7) + 5(0) = 105 + 0 = \$105$

Hence, vertex (7, 0) gives a maximum profit of $105.

15. $P = 120x + 340y$

(0, 0) $P = 120(0) + 340(0) = 0 + 0 = \$0$

(20, 50) $P = 120(20) + 340(50) = 2400 + 17{,}000$
               $= \$19{,}400$

(0, 62) $P = 120(0) + 340(62) = 0 + 21{,}080$
               $= \$21{,}080$

(43, 0) $P = 120(43) + 340(0) = 5160 + 0$
               $= \$5160$

Hence, vertex (0, 62) gives a maximum profit of $21,080.

17. Since the profit on the sale of a television is $40.00 and the profit on the sale of a VCR is $60.00, the objective function is

$P = 40t + 60v$

where t = number of televisions sold and v = number of VCRs sold.

Write the system of constraints.

$t + v \leq 30$

   $t \geq 2v$

   $t \geq 0$

   $v \geq 0$

Graph the linear system and find the vertices.

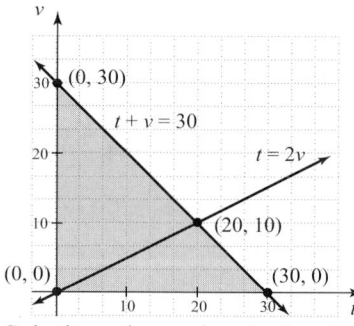

Substitute the vertices in the objective function and find the maximum value.

$P = 40t + 60v$

(0, 0) $P = 40(0) + 60(0) = 0 + 0 = 0$

(20, 10) $P = 40(20) + 60(10) = 800 + 600$
               $= 1400$

(30, 0) $P = 40(30) + 60(0) = 1200 + 0 = 1200$

(0, 30) $P = 40(0) + 60(30) = 0 + 1800 = 1800$

Hence, vertex (0, 30) gives a maximum profit of $1800. The owner should stock 0 television sets and 30 VCRs.

19. Since the profit on each rabbit is \$5.00 and the profit on each chick is \$3.00, the objective function is
$P = 5r + 3c$
where r = number of rabbits sold and c = number of chicks sold.
Write the system of constraints.
$$r + c \leq 50$$
$$4r + 3c \leq 360$$
$$r \geq 0$$
$$c \geq 0$$
Graph the linear system and find the vertices.

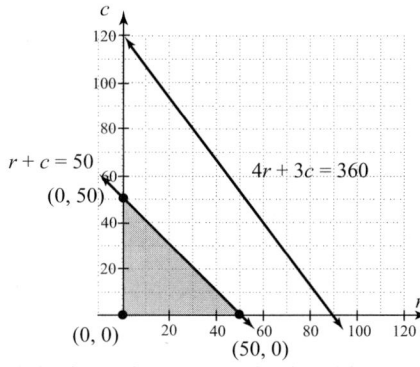

Substitute the vertices in the objective function and find the maximum value.
$P = 5r + 3c$
$(0, 0) \quad P = 5(0) + 3(0) = 0 + 0 = 0$
$(0, 50) \quad P = 5(0) + 3(50) = 0 + 150 = 150$
$(50, 0) \quad P = 5(50) + 3(0) = 250 + 0 = 250$
Hence, the vertex $(50, 0)$ gives a maximum profit of \$250. The person should raise 50 rabbits and 0 chicks.

21. Since the profit for beans is \$65.00 per acre and the profit for corn is \$40.00 per acre, the objective function is
$P = 65b + 40c$
where b = acres of beans grown and c = acres of corn grown.
Write the system of constraints.
$$b + c \leq 50$$
$$40b + 30c \leq 3200$$
$$b \geq 0$$
$$c \geq 0$$
Graph the linear system and find the vertices.

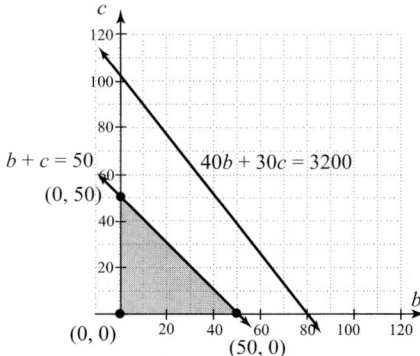

Substitute the vertices in the objective function and find the maximum value.

$P = 65b + 40c$
$(0, 0) \quad P = 65(0) + 40(0) = 0 + 0 = 0$
$(50, 0) \quad P = 65(50) + 40(0) = 3250 + 0 = 3250$
$(0, 50) \quad P = 65(0) + 40(50) = 0 + 2000 = 2000$
Hence, the vertex $(50, 0)$ gives a maximum profit of \$3250. The owner should grow 50 acres of beans and 0 acres of corn.

23. Linear programming is used in making decisions and in finding cost effective solutions. A common problem is to find "the most..." or "the least...."

25. It's a function that relates the variables in linear programming.

27. Evaluate the objective function at the coordinates of each vertex of the region. From the various results, choose the maximum or minimum, as desired.

Exercise Set 8-5

1. $\{(5, 8), (6, 9), (7, 10), (8, 11)\}$
Domain = $\{5, 6, 7, 8\}$; range = $\{8, 9, 10, 11\}$
The relation is a function.

3. $\{(6, 11), (7, 11), (8, 11), (9, 11)\}$
Domain = $\{6, 7, 8, 9\}$; range = $\{11\}$
The relation is a function.

5. $\{(0, 0)\}$
Domain = $\{0\}$; range = $\{0\}$
The relation is a function. (The requirement for a function is trivially or automatically satisfied.)

7. $\{(-10, 20), (-10, 40), (-40, 60), (-60, 60)\}$
Domain = $\{-10, -40, -60\}$; range = $\{20, 40, 60\}$
The relation is not a function since -10 is paired with more than one element.

9. $f(x) = 3x + 8$
$f(3) = 3(3) + 8$
$f(3) = 9 + 8$
$f(3) = 17$

11. $f(x) = 4x - 8$
$f(-10) = 4(-10) - 8$
$f(-10) = -40 - 8$
$f(-10) = -48$

13. $f(x) = 8x^2 + 3x$
$f(0) = 8(0)^2 + 3(0)$
$f(0) = 0 + 0$
$f(0) = 0$

15. $f(x) = x^2 + 4x + 7$

$f(-3.6) = (-3.6)^2 + 4(-3.6) + 7$
$f(-3.6) = 12.96 - 14.4 + 7$
$f(-3.6) = 5.56$

17. Select $x = 0$, $x = 1$, and $x = 2$ and find $f(x)$ for each x.

$f(0) = 7(0) - 8$ $f(1) = 7(1) - 8$ $f(2) = 7(2) - 8$
$f(0) = 0 - 8$ $f(1) = 7 - 8$ $f(2) = 14 - 8$
$f(0) = -8$ $f(1) = -1$ $f(2) = 6$

Plot the points $(0, -8)$, $(1, -1)$, and $(2, 6)$ on the graph and draw a line through them.

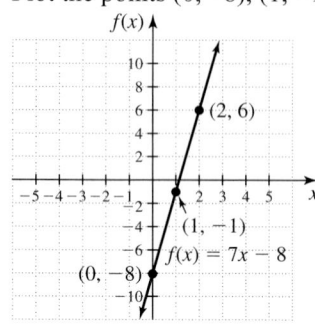

19. Select $x = 0$, $x = 1$, and $x = 2$ and find $f(x)$ for each value of x.

$f(0) = -3(0) + 2$ $f(1) = -3(1) + 2$ $f(2) = -3(2) + 2$
$f(0) = 0 + 2$ $f(1) = -3 + 2$ $f(2) = -6 + 2$
$f(0) = 2$ $f(1) = -1$ $f(2) = -4$

Plot the points $(0, 2)$, $(1, -1)$, and $(2, -4)$ on the graph and draw a line through them.

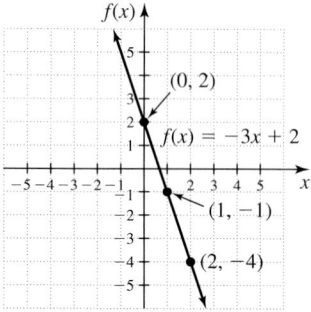

21. For $y = x^2$, $a = 1$, $b = 0$, and $c = 0$.

Find the vertex.

$$x = \frac{-b}{2a} = \frac{-0}{2(1)} = 0$$

$y = 0^2 = 0$

The vertex is $(0, 0)$.
Find the y intercept.

$y = 0^2 = 0$

The y intercept is $(0, 0)$.
Find the x intercept.

$0 = x^2$
$0 = x$

The x intercept is $(0, 0)$ as well.
Select values for x to find points on the parabola.

x	$y = x^2$	(x, y)
-3	$y = (-3)^2 = 9$	$(-3, 9)$
-2	$y = (-2)^2 = 4$	$(-2, 4)$
-1	$y = (-1)^2 = 1$	$(-1, 1)$
1	$y = 1^2 = 1$	$(1, 1)$
2	$y = 2^2 = 4$	$(2, 4)$
3	$y = 3^2 = 9$	$(3, 9)$

The parabola opens upward because $a > 0$.

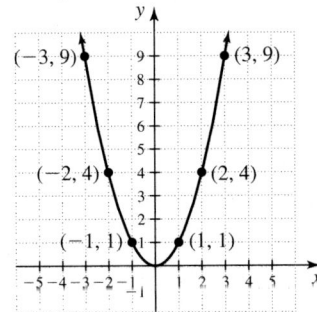

23. For $y = x^2 + 6x + 9$, $a = 1$, $b = 6$, and $c = 9$.

Find the vertex.

$$x = \frac{-b}{2a} = \frac{-6}{2(1)} = -3$$

$$y = (-3)^2 + 6(-3) + 9 = 9 - 18 + 9 = 0$$

The vertex is $(-3, 0)$.

Find the y intercept.

$$y = 0^2 + 6(0) + 9 = 9$$

The y intercept is $(0, 9)$.

Find the x intercept.

$$0 = x^2 + 6x + 9$$
$$0 = (x + 3)^2$$
$$0 = x + 3$$
$$-3 = x$$

The x intercept is $(-3, 0)$.

Select values for x to find points on the parabola.

x	$y = x^2 + 6x + 9$	(x, y)
-6	$y = (-6)^2 + 6(-6) + 9 = 36 - 36 + 9 = 9$	$(-6, 9)$
-5	$y = (-5)^2 + 6(-5) + 9 = 25 - 30 + 9 = 4$	$(-5, 4)$
-1	$y = (-1)^2 + 6(-1) + 9 = 1 - 6 + 9 = 4$	$(-1, 4)$

The parabola opens upward because $a > 0$.

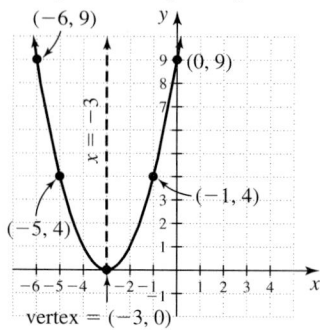

25. For $y = -x^2 + 12x - 36$, $a = -1$, $b = 12$, and $c = -36$.
Find the vertex.
$$x = \frac{-b}{2a} = \frac{-12}{2(-1)} = 6$$
$$y = -6^2 + 12(6) - 36 = -36 + 72 - 36 = 0$$
The vertex is $(6, 0)$.
Find the y intercept.
$$y = -0^2 + 12(0) - 36 = -36$$
The y intercept is $(0, -36)$.
Find the x intercept.
$$0 = -x^2 + 12x - 36$$
$$0 = -1(x^2 - 12x + 36)$$
$$0 = -1(x - 6)^2$$
$$0 = x - 6$$
$$6 = x$$
The x intercept is $(6, 0)$.
Select values for x to find points on the parabola.

x	$y = -x^2 + 12x - 36$	(x, y)
3	$y = -3^2 + 12(3) - 36 = -9$	$(3, -9)$
4	$y = -4^2 + 12(4) - 36 = -4$	$(4, -4)$
8	$y = -8^2 + 12(8) - 36 = -4$	$(8, -4)$
9	$y = -9^2 + 12(9) - 36 = -9$	$(9, -9)$

The parabola opens downward because $a < 0$.

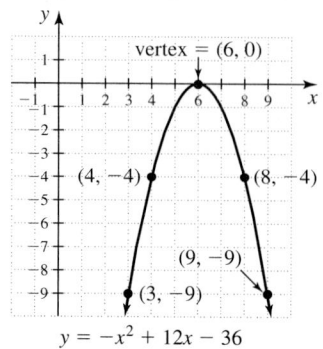

27. For $y = -10x^2 + 20x$, $a = -10$, $b = 20$, and $c = 0$.
Find the vertex.
$$x = \frac{-b}{2a} = \frac{-20}{2(-10)} = 1$$
$$y = -10(1)^2 + 20(1) = 10$$
The vertex is $(1, 10)$.
Find the y intercept.
$$y = -10(0)^2 + 20(0) = 0$$
The y intercept is $(0, 0)$.
Find the x intercepts.
$$0 = -10x^2 + 20x$$
$$0 = x(-10x + 20)$$
$$0 = x \quad \begin{array}{l} 0 = -10x + 20 \\ -20 = -10x \\ 2 = x \end{array}$$
The x intercepts are $(0, 0)$ and $(2, 0)$.
Select values for x to find points on the parabola.

x	$y = -10x^2 + 20x$	(x, y)
$\frac{1}{2}$	$y = -10\left(\frac{1}{2}\right)^2 + 20\left(\frac{1}{2}\right) = 7\frac{1}{2}$	$\left(\frac{1}{2}, 7\frac{1}{2}\right)$
$\frac{3}{2}$	$y = -10\left(\frac{3}{2}\right)^2 + 20\left(\frac{3}{2}\right) = 7\frac{1}{2}$	$\left(\frac{3}{2}, 7\frac{1}{2}\right)$

The parabola opens downward because $a < 0$.

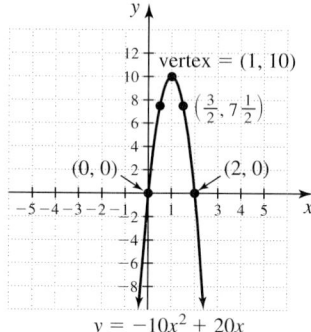

29. For $y = 5^x$, the function increases as x increases since $a > 1$.
Select values for x and find y.

x	$y = 5^x$	(x, y)
-1	$y = 5^{-1} = 0.2$	$(-1, 0.2)$
0	$y = 5^0 = 1$	$(0, 1)$
1	$y = 5^1 = 5$	$(1, 5)$

Plot the points and connect with a smooth curve.

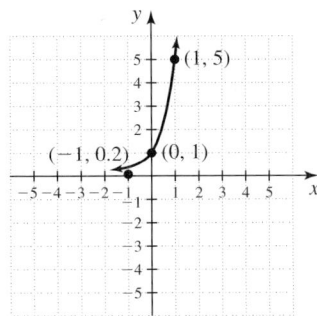

31. For $y = \left(\frac{1}{2}\right)^{x-2}$, the function decreases as x

increases since $0 < a < 1$.
Select values for x and find y.

x	$x = \left(\frac{1}{2}\right)^{x-2}$	(x, y)
-1	$y = \left(\frac{1}{2}\right)^{-1-2} = 8$	$(-1, 8)$
0	$y = \left(\frac{1}{2}\right)^{0-2} = 4$	$(0, 4)$
1	$y = \left(\frac{1}{2}\right)^{1-2} = 2$	$(1, 2)$
2	$y = \left(\frac{1}{2}\right)^{2-2} = 1$	$(2, 1)$

Plot the points and connect with a smooth curve.

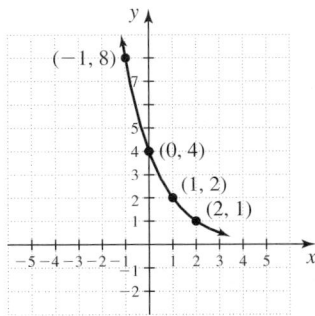

33. Since no vertical line can intersect the graph at more than one point, the relation is a function.

35. Since a vertical line can intersect the graph at more than one point, the relation is not a function.

37. Since no vertical line can intersect the graph at more than one point, the relation is a function.

39. For $f(t) = -16t^2 + 60t + 6$, $a = -16$, $b = 60$, and $c = 6$.
To find the maximum height, find $f(t)$ at the vertex.

$$t = \frac{-b}{2a} = \frac{-60}{2(-16)} = 1.875$$

$f(1.875) = -16(1.875)^2 + 60(1.875) + 6 = 62.25$
The maximum height is 62.25 feet.
To find the duration of the entire flight, find the value of t at the right-most t intercept.

$$0 = -16t^2 + 60t + 6$$

$$t = \frac{-b \pm \sqrt{b^2 - 4ac}}{2a}$$

$$t = \frac{-60 \pm \sqrt{60^2 - 4(-16)(6)}}{2(-16)}$$

$t = -0.10$ or $t = 3.85$
The duration of the entire flight is 3.85 seconds.

41. The function $y = x(16 - 2x)$ or $y = -2x^2 + 16x$ describes the capacity of the gutter. To find the maximum capacity, find the x coordinate of the vertex.

$$x = \frac{-b}{2a} = \frac{-16}{2(-2)} = 4$$

The gutter has a base of $16 - 2(4)$ or 8 inches and folded sides of 4 inches each.

43. Let $t = 0.5$ $\left(5 \text{ years is } \frac{1}{2} \text{ of a decade}\right)$ and

$A_0 = 4,000,000$.

$f(t) = A_0(1.4)^t$

$f(0.5) = 4,000,000(1.4)^{0.5}$
$f(0.5) \approx 4,732,864$
The population in 5 years is 4,732,864.

45. Let $A_0 = 5$ and $x = 20$.

$F(x) = A_0 2^{-0.23x}$
$F(20) = (5)2^{-0.23(20)}$
$F(20) \approx 0.206$
There will be 0.206 milligram after 20 days.

47. A function is a special relation; more specifically: a relation in which there exists one and only one second element for each first element in the ordered pairs.

49. The set of second elements of the ordered pairs is called the range.

51. A parabola is a U-shaped curve, which is the graph of a quadratic function.

53. Have the equation in the form
$f(x)[\text{or } y] = ax^2 + bx + c$, where a, b, and c, are constants. If $a > 0$, then the parabola opens upward.

55. (a) Compound interest. This occurs when interest on an account is not collected but kept, and thus contributes to the earnings.

(b) Radioactive decay; the total amount (of a substance) lost depends on the temporary amount present "along the way."

57. Find the vertex.

$$y = \frac{-2}{2(-1)} = 1$$

$$x = -1^2 + 2(1) + 4 = 5$$

The vertex is (5, 1).
Find the x intercept.

$$x = -0^2 + 2(0) + 4 = 4$$

The x intercept is (4, 0).
Find the y intercepts.

$$0 = -y^2 + 2y + 4$$

$$y = \frac{-2 \pm \sqrt{2^2 - 4(-1)(4)}}{2(-1)}$$

$$y = 1 \pm \sqrt{5}$$

The y intercepts are $\left(0, 1 + \sqrt{5}\right)$ and $\left(0, 1 - \sqrt{5}\right)$.

The graph opens to the left.

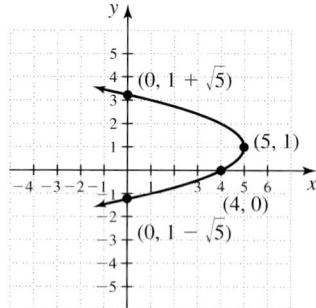

Review Exercises

1. Substitute $x = 0$, $x = 1$, and $x = 2$ in the equation $4x - y = 8$ and solve each equation for y. The three points are $(0, -8)$, $(1, -4)$, and $(2, 0)$. Plot the points on the plane and draw a straight line through them.

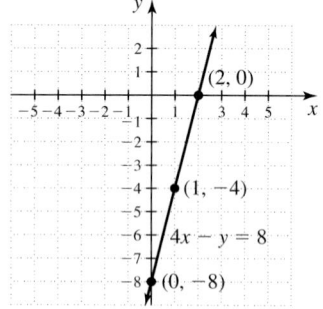

3. The graph of $x = -5$ is a vertical line passing through $(-5, 0)$ on the x axis.

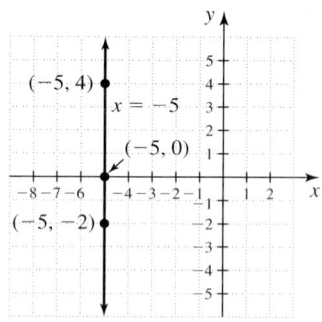

5. Substitute $x = 0$, $x = 1$, and $x = 2$ in the equation $y = -3x + 10$ and solve each equation for y. The three points are $(0, 10)$, $(1, 7)$, and $(2, 4)$. Plot the points on the plane and draw a straight line through them.

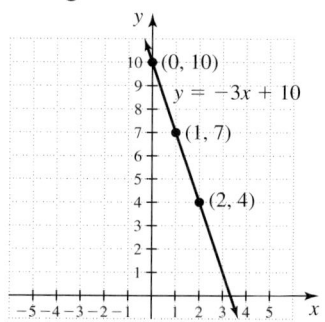

7. $m = \dfrac{y_2 - y_1}{x_2 - x_1} = \dfrac{3 - 6}{4 - (-8)} = \dfrac{-3}{12} = -\dfrac{1}{4}$

The slope is $-\dfrac{1}{4}$.

9. $m = \dfrac{y_2 - y_1}{x_2 - x_1} = \dfrac{8 - (-6)}{-3 - (-5)} = \dfrac{14}{2} = 7$

The slope is 7.

11. $m = \dfrac{y_2 - y_1}{x_2 - x_1} = \dfrac{9 - 9}{-3 - 5} = \dfrac{0}{-8} = 0$

The slope is 0.

13. Solve the equation for y.

$$3x + y = 12$$
$$y = -3x + 12$$

The slope $= -3$ and the y intercept $= (0, 12)$.
To find the x intercept, let $y = 0$ and solve for x.

$$3x + 0 = 12$$
$$3x = 12$$
$$x = 4$$

The x intercept $= (4, 0)$.

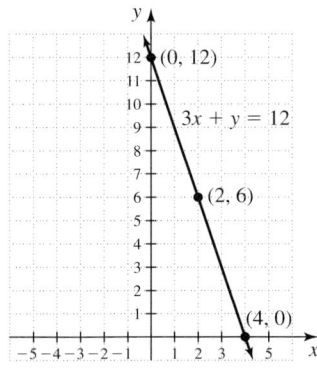

15. Solve the equation for y.
$$4x - 7y = 28$$
$$-7y = -4x + 28$$
$$y = \frac{4}{7}x - 4$$

The slope $= \frac{4}{7}$ and the y intercept $= (0, -4)$.

To find the x intercept, let $y = 0$ and solve for x.
$$4x - 7(0) = 28$$
$$4x = 28$$
$$x = 7$$
The x intercept $= (7, 0)$.

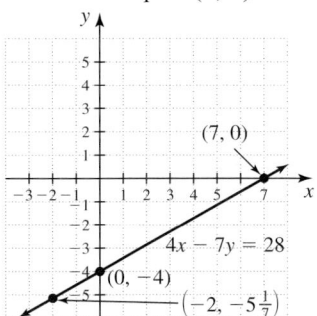

17. Let x = amount Betty has.
Then $x + 16.72$ = amount Mary has.
Together they have $32.00.
$$x + x + 16.72 = 32$$
$$2x + 16.72 = 32$$
$$2x = 15.28$$
$$x = 7.64$$
$$x + 16.72 = 24.36$$
Betty has $7.64 and Mary has $24.36.

19. Draw the graphs for both equations on the Cartesian plane.

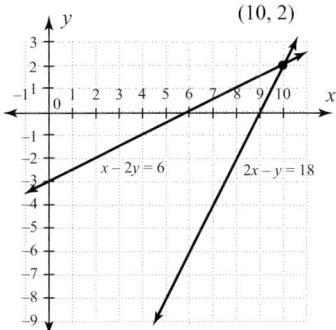

The point of intersection is (10, 2). Hence, the solution is $\{(10, 2)\}$.

21. Draw the graphs for both equations on the Cartesian plane.

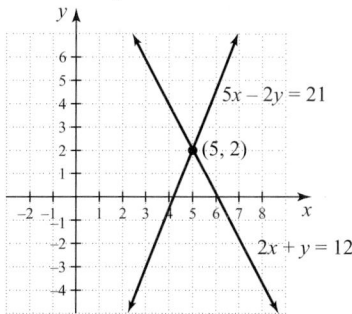

The point of intersection is (5, 2). Hence, the solution is $\{(5, 2)\}$.

23. Select one equation and solve for one variable.
$$x - y = 11$$
$$x = y + 11$$

Substitute $x = y + 11$ in the first equation and solve for y.
$$x + 6y = 4$$
$$y + 11 + 6y = 4$$
$$7y = -7$$
$$y = -1$$

Substitute $y = -1$ in one equation and solve for x.
$$x - y = 11$$
$$x - (-1) = 11$$
$$x = 10$$
The solution set is $\{(10, -1)\}$.

25. Select one equation and solve for one variable.
$$x - y = 3$$
$$x = y + 3$$

Substitute $x = y + 3$ in the second equation and solve for y.
$$x - 3y = -3$$
$$y + 3 - 3y = -3$$
$$-2y = -6$$
$$y = 3$$

Substitute $y = 3$ in one equation and solve for x.
$$x - y = 3$$
$$x - 3 = 3$$
$$x = 6$$
The solution set is $\{(6, 3)\}$.

27. Multiply the first equation by 2. Then add the equations to eliminate the y variable.
$$6x - 2y = 24$$
$$x + 2y = 4$$
$$\overline{7x \qquad = 28}$$
$$x = 4$$

Substitute $x = 4$ in one equation and solve for y.

$$3x - y = 12$$
$$3(4) - y = 12$$
$$12 - y = 12$$
$$-y = 0$$
$$y = 0$$

The solution set is $\{(4, 0)\}$.

29. Multiply the second equation by -2 and add the equations to eliminate the y variable.

$$
\begin{array}{rcl}
4x - 2y &=& 11 \\
-10x + 2y &=& -20 \\
\hline
-6x &=& -9
\end{array}
$$

$$x = \frac{3}{2}$$

Substitute $x = \frac{3}{2}$ in one equation and solve for y.

$$4x - 2y = 11$$
$$4\left(\frac{3}{2}\right) - 2y = 11$$
$$6 - 2y = 11$$
$$-2y = 5$$
$$y = -\frac{5}{2}$$

The solution set is $\left\{\left(\frac{3}{2}, -\frac{5}{2}\right)\right\}$.

31. Let x = cost per pound of coffee and y = cost per pound of tea.
The system is
$$3x + 4y = 14.50$$
$$5x + 2y = 16.00$$

Multiply the second equation by -2. Add the equations to eliminate the y variable.

$$
\begin{array}{rcl}
3x + 4y &=& 14.5 \\
-10x - 4y &=& -32 \\
\hline
-7x &=& -17.5 \\
x &=& 2.5
\end{array}
$$

Substitute $x = 2.5$ in one equation and solve for y.

$$3x + 4y = 14.5$$
$$3(2.5) + 4y = 14.5$$
$$7.5 + 4y = 14.5$$
$$4y = 7$$
$$y = 1.75$$

The coffee is $2.50 per pound and the tea is $1.75 per pound.

33. Graph the solid (since the inequality sign is $\geq$) line $x - 19y = 5$ using two points, such as $\left(0, -\frac{5}{19}\right)$ and $(5, 0)$. Select $(0, 0)$ as a test point.

$$x - 19y \geq 5$$
$$0 - 19(0) \geq 5?$$
$$0 \geq 5 \text{ False}$$

Since this is a false statement, shade the area that does not contain the test point.

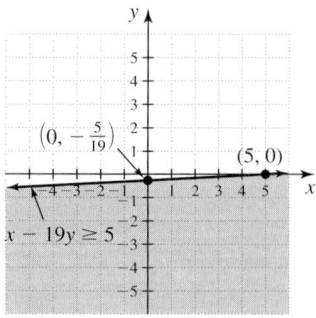

35. Graph the solid (since the inequality sign is $\leq$) line $-3x + 10y \leq -15$ using two points such as $\left(0, -\frac{3}{2}\right)$ and $(5, 0)$. Select $(0, 0)$ as a test point.

$$-3x + 10y \leq -15$$
$$-3(0) + 10(0) \leq -15?$$
$$0 \leq -15 \text{ False}$$

Since this is a false statement, shade the area that does not contain the test point.

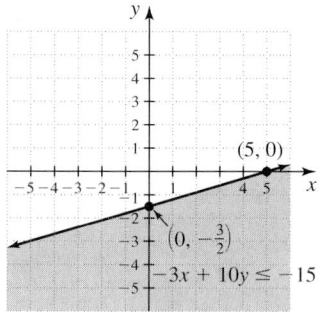

37. Graph $5x - y > 10$ using a dashed line (since the inequality sign is $>$) and two points, such as $(2, 0)$ and $(0, -10)$. Select $(0, 0)$ as a test point.
$$5x - y > 10$$
$$5(0) - 0 > 10?$$
$$0 > 10 \text{ False}$$
Since this is a false statement, shade the area that does not contain the test point.
Graph $2x + 3y \leq 12$ using a solid line (since the inequality sign is $\leq$) and the two points, such as $(3, 2)$ and $(6, 0)$. Select $(0, 0)$ as a test point.
$$2x + 3y \leq 12$$
$$2(0) + 3(0) \leq 12?$$
$$0 \leq 12 \text{ True}$$
Since this is a true statement, shade the area that contains the test point.
The solution is the intersection of the two half planes and part of the line $2x + 3y = 12$.

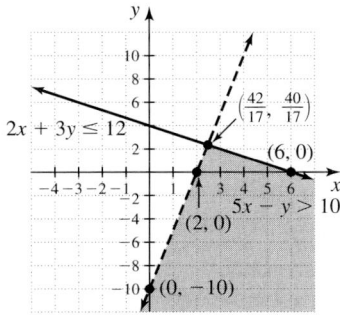

$2x + 3y \leq 12$

$\left(\frac{42}{17}, \frac{40}{17}\right)$

$(6, 0)$

$5x - y > 10$

$(2, 0)$

$(0, -10)$

39. Graph $12x + 3y \geq 24$ using a solid line (since the inequality sign is $\geq$) and two points, such as (2, 0) and (0, 8). Select (0, 0) as a test point.

$$12x + 3y \geq 24$$
$$12(0) + 3(0) \geq 24?$$
$$0 \geq 24 \text{ False}$$

Since this is a false statement, shade the area that does not contain the test point.

Graph $-x + y \leq 5$ using a solid line (since the inequality sign is $\leq$) and two points, such as (−5, 0) and (0, 5). Select (0, 0) as a test point.

$$-x + y \leq 5$$
$$0 + 0 \leq 5?$$
$$0 \leq 5 \text{ True}$$

Since this is a true statement, shade the area containing the test point.

The solution is the intersection of the two half planes and part of each line $12x + 3y = 24$ and $-x + y = 5$.

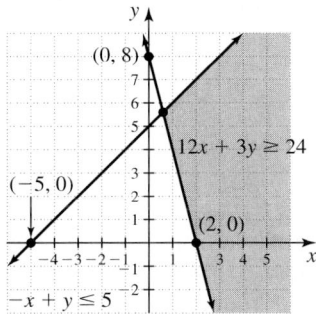

$(0, 8)$

$12x + 3y \geq 24$

$(-5, 0)$

$(2, 0)$

$-x + y \leq 5$

41. Since the profit on Model A is $20.00 and the profit on Model B is $10.00, the objective function is

$$P = 20a + 10b$$

where a = number of Model A and b = number of Model B.

Write the system of constraints.

$$6a + 5b \leq 160$$
$$2a + b \leq 100$$
$$a \geq 0$$
$$b \geq 0$$

Graph the linear system and find the vertices.

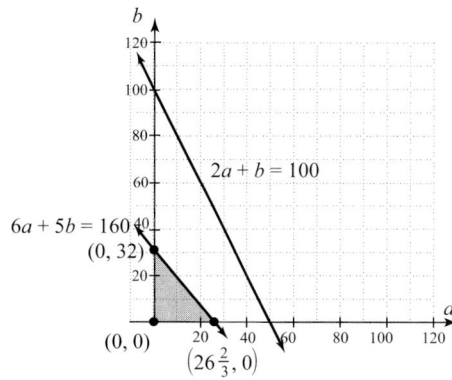

$2a + b = 100$

$6a + 5b = 160$

$(0, 32)$

$(0, 0)$

$\left(26\frac{2}{3}, 0\right)$

Substitute the vertices in the objective function and find the maximum value. The point (26, 0) is the closest to $\left(26\frac{2}{3}, 0\right)$ and still maintains whole numbers.

$P = 20a + 10b$
(0, 32) $P = 20(0) + 10(32) = 0 + 320 = 320$
(0, 0) $P = 20(0) + 10(0) = 0 + 0 = 0$
(26, 0) $P = 20(26) + 10(0) = 520 + 0 = 520$

Hence, point (26, 0) gives a maximum profit of $520.00. The manufacturer should produce 26 of model A.

43. {(2, 5), (5, −7), (6, −10)}
Domain = {2, 5, 6}; range = {5, −7, −10}
The relation is a function.

45. {(4, 10), (5, 10), (6, 10)}
Domain = {4, 5, 6}; range = {10}
The relation is a function.

47. $f(x) = 5x - 12$
$f(5) = 5(5) - 12$
$f(5) = 25 - 12$
$f(5) = 13$

49. $f(x) = x^2 + 7x + 10$
$f(-10) = (-10)^2 + 7(-10) + 10$
$f(-10) = 100 - 70 + 10$
$f(-10) = 40$

51. For $y = x^2 + 10x + 25$, $a = 1$, $b = 10$, and $c = 25$.
Find the vertex.

$$x = \frac{-b}{2a} = \frac{-10}{2(1)} = -5$$

$$y = (-5)^2 + 10(-5) + 25 = 0$$

The vertex is (−5, 0).
Find the y intercept.

$$y = 0^2 + 10(0) + 25 = 25$$

The y intercept is (0, 25).
Find the x intercept.

$0 = x^2 + 10x + 25$
$0 = (x+5)^2$
$0 = x+5$
$-5 = x$

The x intercept is $(-5, 0)$.
Select values for x to find points on the parabola.

x	$y = x^2 + 10x + 25$	(x, y)
-7	$y = (-7)^2 + 10(-7) + 25 = 4$	$(-7, 4)$
-6	$y = (-6)^2 + 10(-6) + 25 = 1$	$(-6, 1)$
-4	$y = (-4)^2 + 10(-4) + 25 = 1$	$(-4, 1)$
-3	$y = (-3)^2 + 10(-3) + 25 = 4$	$(-3, 4)$

The parabola opens upward because $a > 0$.

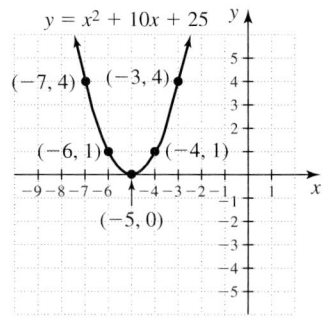

53. Find $y = -x^2 + 25$, $a = -1$, $b = 0$, and $c = 25$.
Find the vertex.

$$x = \frac{-b}{2a} = \frac{-0}{2(-1)} = 0$$

$y = -0^2 + 25 = 25$
The vertex is $(0, 25)$.
Find the y intercept.

$y = -0^2 + 25 = 25$
The y intercept is $(0, 25)$.
Find the x intercepts.

$0 = -x^2 + 25$
$0 = -1(x-5)(x+5)$

$0 = x-5$	$0 = x+5$
$5 = x$	$-5 = x$

The x intercepts are $(-5, 0)$ and $(5, 0)$.
Select values for x to find points on the parabola.

x	$y = -x^2 + 25$	(x, y)
-3	$y = (-3)^2 + 25 = 16$	$(-3, 16)$
3	$y = -(3)^2 + 25 = 16$	$(3, 16)$

The parabola opens downward because $a < 0$.

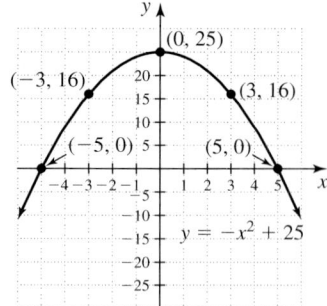

55. Select values for x and find y.

x	$y = -3^x$	(x, y)
0	$y = -3^0 = -1$	$(0, -1)$
1	$y = -3^1 = -3$	$(1, -3)$
$\frac{3}{2}$	$y = -3^{\frac{3}{2}} = -5.2$	$\left(\frac{3}{2}, -5.2\right)$

Plot the points and connect with a smooth curve.

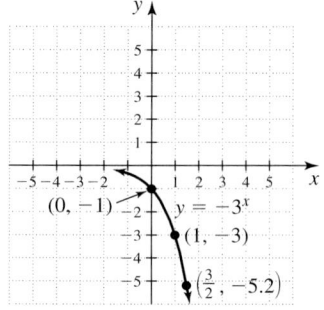

57. Let $x =$ the number of seconds the ball is in the air. The ball is thrown from a height of 4 feet with a velocity of 80 feet per second, so the height of the ball above the ground at time x is given by the function $f(x) = -16x^2 + 80x + 4$.
The maximum height the ball can attain is the y coordinate of the vertex of the parabola determined by this function. The x coordinate of the vertex is

$$x = -\frac{b}{2a} = -\frac{80}{2(-16)} = \frac{80}{32} = 2.5$$

The y coordinate of the vertex is

$y = -16(2.5)^2 + 80(2.5) + 4$
$\quad = -16(6.25) + 80(2.5) + 4$
$\quad = -100 + 200 + 4$
$\quad = 104$

Therefore the ball is 104 feet above the ground at its maximum height. That is 100 feet above the point of launch.

Chapter Test

1. Substitute $x = -5$ and $x = 0$ in the equation $3x - 5y = -15$ and solve each equation for y. The two points are $(-5, 0)$ and $(0, 3)$. Plot the points on the plane and draw a straight line through them.

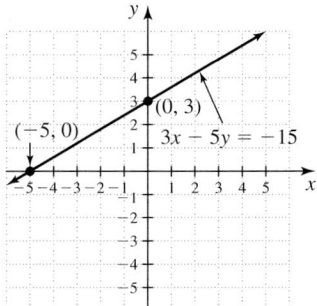

3. The graph of $x = 0$ is a vertical line passing through $(0, 0)$ on the x axis.

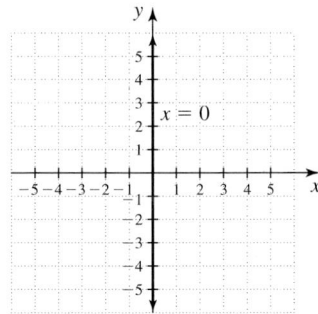

5. Substitute $x = -4$, $x = 0$, and $x = 2$ in the equation $2x + 3y = -8$ and solve each equation for y. The three points are $(-4, 0)$, $\left(0, -2\frac{2}{3}\right)$ and $(2, -4)$.

Plot the points on the plane and draw a straight line through them.

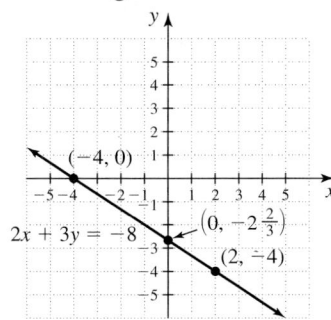

7. $m = \dfrac{y_2 - y_1}{x_2 - x_1} = \dfrac{7 - (-10)}{2 - (-5)} = \dfrac{17}{7}$

The slope is $\dfrac{17}{7}$.

9. Solve the equation for y.
$$x + 5y = 20$$
$$5y = -x + 20$$
$$y = -\frac{x}{5} + 4$$

The slope $= -\dfrac{1}{5}$ and the y intercept $= (0, 4)$.

To find the x intercept, let $y = 0$ and solve for x.
$$x + 5(0) = 20$$
$$x = 20$$
The x intercept $= (20, 0)$.

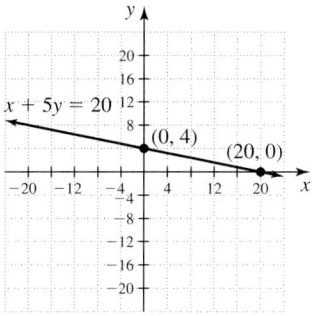

11. Multiply the second equation by 7 and add the equations to eliminate the y variable.
$$\begin{aligned} x - 7y &= 10 \\ 14x + 7y &= 35 \\ \hline 15x &= 45 \\ x &= 3 \end{aligned}$$

Substitute $x = 3$ in one equation and solve for y.
$$2x + y = 5$$
$$2(3) + y = 5$$
$$6 + y = 5$$
$$y = -1$$

The solution set is $\{(3, -1)\}$.

13. Multiply the first equation by -3 and add the equations.
$$\begin{aligned} -9x - 12y &= -57 \\ 9x + 12y &= 57 \\ \hline 0 &= 0 \end{aligned}$$

Since the variables are eliminated and $0 = 0$ is true, the system is dependent. The solution is $\{(x, y) \mid 3x + 4y = 19\}$.

15. Graph $x + 6y \geq 10$ using a solid line (since the inequality sign is $\geq$) and two points, such as $\left(0, \dfrac{5}{3}\right)$ and $(10, 0)$. Select $(0, 0)$ as a test point.
$$x + 6y \geq 10$$
$$0 + 6(0) \geq 10 ?$$
$$0 \geq 10 \text{ False}$$
Since this is a false statement, shade the area that does not contain the test point.
Graph $6x - y \leq 23$ using a solid line (since the inequality sign is $\leq$) and two points, such as $(5, 7)$ and $(4, 1)$. Select $(0, 0)$ as a test point.

$6x - y \le 23$
$6(0) - 0 \le 23$?
$\qquad 0 \le 23$ True
Since this is a true statement, shade the area that contains the test point.
The solution is the intersection of the two half planes and parts of the lines $x + 6y = 10$ and $6x - y = 23$.

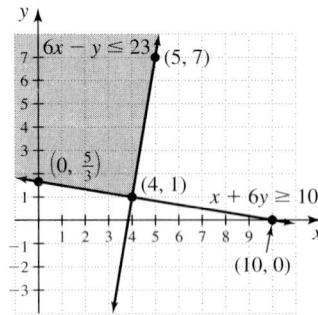

17. $\{(-4, 6), (-10, 18), (12, 5)\}$
Domain = $\{-4, -10, 12\}$; range = $\{6, 18, 5\}$
The relation is a function.

19. $\quad f(x) = -3x + 10$
$\quad f(-15) = -3(-15) + 10$
$\quad f(-15) = 45 + 10$
$\quad f(-15) = 55$

21. For $y = x^2 - 9x + 14$, $a = 1$, $b = -9$, and $c = 14$.
Find the vertex.
$$x = \frac{-b}{2a} = \frac{-(-9)}{2(1)} = \frac{9}{2} \text{ or } 4\frac{1}{2}$$
$$y = \left(\frac{9}{2}\right)^2 - 9\left(\frac{9}{2}\right) + 14 = -\frac{25}{4} \text{ or } -6\frac{1}{4}$$
The vertex is $\left(4\frac{1}{2}, -6\frac{1}{4}\right)$.

Find the y intercept.
$y = 0^2 - 9(0) + 14 = 14$
The y intercept is $(0, 14)$.
Find the x intercept.
$0 = x^2 - 9x + 14$
$0 = (x - 2)(x - 7)$
$0 = x - 2 \quad | \quad 0 = x - 7$
$x = 2 \qquad | \quad x = 7$
The x intercepts are $(2, 0)$ and $(7, 0)$.
Select values for x to find points on the parabola.

x	$y = x^2 - 9x + 14$	(x, y)
3	$y = 3^2 - 9(3) + 14 = -4$	$(3, -4)$
6	$y = 6^2 - 9(6) + 14 = -4$	$(6, -4)$

The parabola opens upward because $a > 0$.

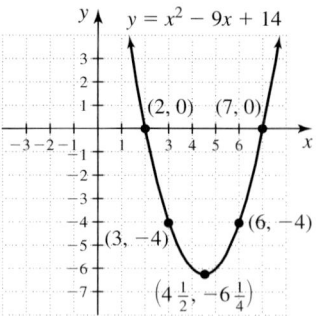

23. Select values for x and find y.

x	$y = 2^{1.5x}$	(x, y)
-1	$y = 2^{1.5(-1)} = 0.35$	$(-1, 0.35)$
0	$y = 2^{1.5(0)} = 1$	$(0, 1)$
1	$y = 2^{1.5(1)} = 2.8$	$(1, 2.8)$
$\frac{3}{2}$	$y = 2^{1.5\left(\frac{3}{2}\right)} = 4.8$	$\left(\frac{3}{2}, 4.8\right)$
2	$y = 2^{1.5(2)} = 8$	$(2, 8)$

Plot the points and connect with a smooth curve.

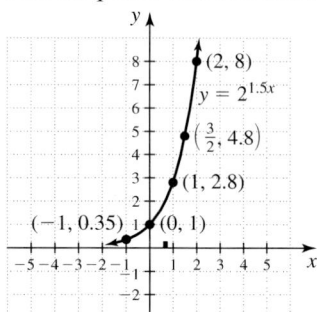

25. Let x = students in one section and let y = students in other section.
Write the system.
$x + y = 57$
$\quad x = y + 5$
Substitute $x = y + 5$ in the first equation.
$\quad x + y = 57$
$y + 5 + y = 57$
$\qquad 2y = 52$
$\qquad y = 26$
Substitute $y = 26$ in one equation and solve for x.
$x = y + 5$
$x = 26 + 5$
$x = 31$
There were 26 and 31 students in each section.

27. Let x = amount earned by cooks and let y = amount earned by servers.
Write the system.
$$3x + 8y = 66.5$$
$$x = y + 2$$

Substitute $x = y + 2$ in the first equation.
$$3x + 8y = 66.5$$
$$3(y + 2) + 8y = 66.5$$
$$3y + 6 + 8y = 66.5$$
$$11y = 60.5$$
$$y = 5.5$$

Substitute $y = 5.5$ in one equation and solve for x.
$$x = y + 2$$
$$x = 5.5 + 2$$
$$x = 7.5$$
The cooks earn \$7.50 per hour and the servers earn \$5.50 per hour.

29. The function $y = x(24 - 2x)$ or $y = -2x^2 + 24x$ describes the capacity of the drain. To find the maximum capacity, find the x coordinate of the vertex.
$$x = \frac{-b}{2a} = \frac{-24}{2(-2)} = 6$$

The drain has a base width of $24 - 2(6)$ or 12 inches and sides with heights of 6 inches each.

9 | Consumer Mathematics

Exercise Set 9-1

1. $0.63 = 63\%$

3. $0.025 = 2.5\%$

5. $1.56 = 156\%$

7. $\dfrac{1}{5} = 0.2 = 20\%$

9. $\dfrac{2}{3} = 0.66\overline{6} = 66.\overline{6}\%$ or 66.7%

11. $1\dfrac{1}{4} = 1.25 = 125\%$

13. $18\% = 18.\% = 0.18$

15. $6\% = 6.\% = 0.06$

17. $62.5\% = 0.625$

19. $320\% = 320.\% = 3.2$

21. $24\% = \dfrac{24}{100} = \dfrac{6}{25}$

23. $9\% = \dfrac{9}{100}$

25. $236\% = \dfrac{236}{100} = 2\dfrac{9}{25}$

27. $\dfrac{1}{2}\% = \dfrac{\frac{1}{2}}{100} = \dfrac{1}{200}$

29. $16\dfrac{2}{3}\% = \dfrac{16\frac{2}{3}}{100} = 16\dfrac{2}{3} \div 100 = \dfrac{50}{3} \cdot \dfrac{1}{100} = \dfrac{50}{300} = \dfrac{1}{6}$

31. Find 5% of $299.99.
$0.05 \times 299.99 = 15$ rounded
$299.99 + 15 = 314.99$
The sales tax is $15 and the total cost is $314.99.

33. Find 6% of $149.99.
$0.06 \times 149.99 = 9.00$ rounded
$149.99 + 9.00 = 158.99$
The sales tax is $9.00 and the total cost is $158.99.

35. Find the decrease.
$999.99 - 399.99 = 600$
Percent decrease $= \dfrac{600}{999.99} \approx 0.60 = 60\%$
The reduction in price is 60%.

37. Original price is $249.99 + 80.00 = 329.99$
Percent decrease $= \dfrac{80.00}{329.99} \approx 0.24 = 24\%$
The reduction in price is 24%.

39. Find 40% of $159.99.
$P = R \times B$
$P = 0.4 \times 159.99$
$P \approx 64.00$
$159.99 - 64.00 = 95.99$
The sale price is $95.99.

41. Find 25% of $199.99.
$P = R \times B$
$P = 0.25 \times 199.99$
$P \approx 50.00$
$199.99 - 50.00 = 149.99$
The grill's sale price is $149.99.

43. $75.36 is 6% of what number?
$P = R \times B$
$75.36 = 0.06 \times B$
$1256 = B$
The item he sold cost $1256.

45. Find 19% of 12,872.
$P = R \times B$
$P = 0.19 \times 12{,}872$
$P \approx 2446$
About 2446 were self-help books.

47. Hundredths, or part of a hundred

49. Change the fraction to a decimal; change this decimal to a percent.

51. Use the formula
Part = (Rate = Percentage) × Base: $P = R \times B$.

  I. Finding a percent of a number: Have R as a decimal and solve for P in $P = R \times B$.

  II. Finding what percent one number is of another: Solve for R in the formula $P = R \times B$; change the answer to a percent.

  III. Finding a number when a percent of it is known: Solve for B in the formula $P = R \times B$, but first change R into a decimal.

53. Next day: Find 30% of $100.00.

$P = R \times B$

$P = 0.3 \times 100$

$P = 30$

The stock is worth $100 − $30 or $70 the next day.

Next week: Find 30% of $70.

$P = R \times B$

$P = 0.3 \times 70$

$P = 21$

After both fluctuations, the stock is worth $70 + $21 or $91, which means you do not break even.

Exercise Set 9-2

1. $I = PRT = (12,000)(0.06)(2) = 1440$

The simple interest is $1440.

3.
$$I = PRT$$
$$360 = (1800)(0.10)T$$
$$\frac{360}{(1800)(0.10)} = \frac{(1800)(0.10)T}{(1800)(0.10)}$$
$$2 = T$$

The time is 2 years.

5.
$$I = PRT$$
$$1290 = (4300)(R)(6)$$
$$\frac{1290}{(4300)(6)} = \frac{(4300)(R)(6)}{(4300)(6)}$$
$$0.05 = R$$

The rate is 5%.

7.
$$I = PRT$$
$$354.6 = P(0.09)(4)$$
$$\frac{354.6}{(0.09)(4)} = \frac{P(0.09)(4)}{(0.09)(4)}$$
$$985 = P$$

The principal is $985.

9.
$$I = PRT$$
$$40 = (500)(R)(2.5)$$
$$\frac{40}{(500)(2.5)} = \frac{(500)(R)(2.5)}{(500)(2.5)}$$
$$0.032 = R$$

the interest rate is 3.2%.

11. $I = PRT$

$= (900)(0.095)(1.5)$

$= 128.25$

The simple interest is $128.25.

13.
$$I = PRT$$
$$514.8 = (660)(0.12)T$$
$$\frac{514.8}{(660)(0.12)} = \frac{(660)(0.12)T}{(660)(0.12)}$$
$$6.5 = T$$

The time is 6.5 years.

15.
$$I = PRT$$
$$8571 = (14,285)(R)(6)$$
$$\frac{8571}{(14,285)(6)} = \frac{(14,285)(R)(6)}{(14,285)(6)}$$
$$0.1 = R$$

The rate is 10%.

17.
$$I = PRT$$
$$156 = (325)(R)(8)$$
$$\frac{156}{(325)(8)} = \frac{(325)(R)(8)}{(325)(8)}$$
$$0.06 = R$$

The rate is 6%.

19.
$$I = PRT$$
$$141.75 = (700)(0.0675)T$$
$$\frac{141.75}{(700)(0.0675)} = \frac{(700)(0.0675)T}{(700)(0.0675)}$$
$$3 = T$$

The time is 3 years.

21. $MV = P\left(1 + \dfrac{r}{n}\right)^{nt} = 825\left(1 + \dfrac{0.04}{1}\right)^{1 \cdot 10} = 1221.20$

$I = 1221.20 - 825 = 396.2$

The interest is $396.20 and the maturity value is $1221.20.

23. $MV = P\left(1 + \dfrac{r}{n}\right)^{nt} = 75\left(1 + \dfrac{0.03}{2}\right)^{2 \cdot 6} = 89.67$

$I = 89.67 - 75 = 14.67$

The interest is $14.67 and the maturity value is $89.67.

25. $MV = P\left(1 + \dfrac{r}{n}\right)^{nt}$

$= 625\left(1 + \dfrac{0.08}{4}\right)^{4 \cdot 12}$

$= 1616.92$

$I = 1616.92 - 625 = 991.92$

The interest is $991.92 and the maturity value is $1616.92.

27. $MV = P\left(1 + \dfrac{r}{n}\right)^{nt}$

$= 1995\left(1 + \dfrac{0.05}{2}\right)^{2 \cdot 6}$

$= 2683.05$

$I = 2683.05 - 1995 = 688.05$

The interest is $688.05 and the maturity value is $2683.05.

29.
$$I = PRT$$
$$4046.4 = (8000)(R)(6)$$
$$\frac{4046.4}{(8000)(6)} = \frac{(8000)(R)(6)}{(8000)(6)}$$
$$0.0843 = R$$

The simple interest is 8.43%.

31.
$$I = PRT$$
$$150 = P(0.12)(0.5)$$
$$\frac{150}{(0.12)(0.5)} = \frac{P(0.12)(0.5)}{(0.12)(0.5)}$$
$$2500 = P$$
The principal is $2500.

33.
$$I = PRT$$
$$1282.5 = (4500)(0.095)T$$
$$\frac{1282.5}{(4500)(0.095)} = \frac{(4500)(0.095)T}{(4500)(0.095)}$$
$$3 = T$$
The term is 3 years.

35.
$$I = PRT$$
$$3170.5 = (9325)(0.08)T$$
$$\frac{3170.5}{(9325)(0.08)} = \frac{(9325)(0.08)T}{(9325)(0.08)}$$
$$4.25 = T$$
The term is 4.25 years or 51 months.

37. $I = PRT = (4300)(0.0975)(5) = 2096.25$
The interest is $2096.25.

39.
$$MV = P\left(1+\frac{r}{n}\right)^{nt}$$
$$= 5000\left(1+\frac{0.09}{4}\right)^{4\cdot10}$$
$$= 12,175.94$$
They will have $12,175.94 in 10 years.

41.
$$MV = P\left(1+\frac{r}{n}\right)^{nt}$$
$$= 60,000\left(1+\frac{0.07}{2}\right)^{2.25}$$
$$= 335,095.61$$
She will have $335,095.61 at age 50.

43. $E = \left(1+\frac{r}{n}\right)^{n} - 1 = \left(1+\frac{0.06}{4}\right)^{4} - 1 = 0.0614$
The effective rate is 6.14%.

45. $E = \left(1+\frac{r}{n}\right)^{n} - 1 = \left(1+\frac{0.065}{4}\right)^{4} - 1 = 0.0666$
The effective rate is 6.66%.

47. $N = 2 \times 4 = 8; \ r = \frac{0.075}{2} = 0.0375$
$$FV = P\left(\frac{(1+R)^{N} - 1}{R}\right)$$
$$= 2250\left(\frac{(1+0.0375)^{8} - 1}{0.0375}\right)$$
$$= 20,548.25$$
The future value of the annuity is $20,548.25.

49. $N = 4 \times 20 = 80; \ R = \frac{0.05}{4} = 0.0125$
$$FV = P\left(\frac{(1+R)^{N} - 1}{R}\right)$$
$$= 200\left(\frac{(1+0.0125)^{80} - 1}{0.0125}\right)$$
$$= 27,223.76$$
The future value of the annuity is $27,223.76.

51. $N = 1 \, T \, 5 = 5; \ R = \frac{0.105}{1} = 0.105$
$$FV = P\left(\frac{(1+R)^{N} - 1}{R}\right)$$
$$= 4000\left(\frac{(1+0.105)^{5} - 1}{0.105}\right)$$
$$= 24,664.64$$
The future of the annuity is $24,664.64.

53. $N = 2 \times 10 = 20; \ R = \frac{0.08}{2} = 0.04$
$$FV = P\left(\frac{(1+R)^{N} - 1}{R}\right)$$
$$= 2000\left(\frac{(1+0.04)^{20} - 1}{0.04}\right)$$
$$= 59,556.16$$
They will have saved $59,556.16 in 10 years.

55. Payment for the use of money is called interest.

57. The *term* of a loan is the time (duration) that the loan is in effect.

59. The effective rate is the simple interest rate that would yield the same maturity value over 1 year as the compound interest rate.

61. Personal loan:
$I = PRT = (10,000)(0.09)(6) = 5400$
Auto loan:
$$I = PRT = (10,000)(0.08)\left(\frac{60}{12}\right) = 4000$$
The personal loan of $10,000 at 9% for 6 years has a higher interest rate of $5400 compared to $4000.

63. new truck loan:
$I = PRT = (25,000)(0.18)(10) = 45,000$
repair loan:
$I = PRT = (18,000)(0.125)(8) = 18,000$
The repair loan of $18,000 at $12\frac{1}{2}$% for 8 years is less expensive.

Exercise Set 9-3

1. $I = PRT = (832.5)(0.02)(1) = 16.65$
The finance charge is $16.65.
New balance $= 832.5 + 675 + 16.65 - 400$
$= 1124.15$
The new balance is $1124.15.

3. $I = PRT = (2364.79)(0.0167)(1) = 39.49$
The finance charge is $39.49.
New balance $= 2364.79 + 39.49 + 1964.32 - 1000$
$= 3368.6$
The new balance is $3368.60.

5. $I = PRT = (986.53)(0.0135)(1) = 13.32$
The finance charge is $13.32.
New balance $= 986.53 + 13.32 + 186.5 - 775$
$= 411.35$
The new balance is $411.35.

7. (a)

Date	Balance	Days	Balance × days
9-1	627.75	$10 - 1 = 9$	5649.75
9-10	$627.75 + 87.95 = 715.70$	$15 - 10 = 5$	3578.50
9-15	$715.70 - 200 = 515.70$	$27 - 15 = 12$	6188.40
9-27	$515.70 + 146.22 = 661.92$	$30 - 27 + 1 = 4$	2647.68
Total		30	18,064.33

$\dfrac{18,064}{30} = 602.14$

The average daily balance is $602.14.

(b) $(602.14)(0.012) = 7.23$
The finance charge is $7.23

(c) $661.92 + 7.23 = 669.15$
The new balance is $669.15.

9. (a)

Date	Balance	Days	Balance × days
6-1	157.95	$5 - 1 = 4$	631.80
6-5	$157.95 + 287.62 = 445.57$	$20 - 5 = 15$	6683.55
6-20	$445.57 - 100 = 345.57$	$30 - 20 + 11$	3801.27
Total		30	11,116.62

$\dfrac{11,116.62}{30} = 370.55$

The average daily balance is $370.55.

(b) $(370.55)(0.014) = 5.19$
The finance charge is $5.19.

(c) $345.57 + 5.19 = 350.76$
The new balance is $350.76.

11. (a)

Date	Balance	Days	Balance × days
7-1	65.00	$2 - 1 = 1$	65.00
7-2	$65 + 720.25 = 785.25$	$8 - 2 = 6$	4711.50
7-8	$785.25 - 500 = 285.25$	$17 - 8 = 9$	2567.25
7-17	$285.25 - 100 = 185.25$	$28 - 17 = 11$	2037.75
7-28	$185.25 + 343.97 = 529.22$	$31 - 28 + 1 = 4$	2116.88
Total		31	11,498.38

$$\frac{11,498.38}{31} = 370.92$$

The average daily balance is $370.92.

(b) $(370.92)(0.011) = 4.08$
The finance charge is $4.08.

(c) $529.22 + 4.08 = 533.30$
The new balance is $533.30.

13. $I = (48 \times 110) - 4000 = 1280$

$$\text{APR} \approx \frac{2NI}{P(T+1)} \approx \frac{2(12)(1280)}{4000(48+1)} \approx 0.1567$$

The annual percentage rate is approximately 15.67%.

15. $I = (72 \times 19.80) - 850 = 575.60$

$$\text{APR} \approx \frac{2NI}{P(T+1)} \approx \frac{2(12)(575.60)}{(850)(72+1)} \approx 0.2226$$

The annual percentage rate is approximately 22.26%.

17. $I = (36 \times 116) - 3256 = 920$

$$\text{APR} \approx \frac{2NI}{P(T+1)} \approx \frac{2(12)(920)}{(3256)(36+1)} \approx 0.1833$$

The annual percentage rate is approximately 18.33%.

19. $f = (36 \times 141.17) - 4200 = 882.12$
$k = 36 - 20 = 16$

$$u = \frac{fk(k+1)}{n(n+1)} = \frac{882.12(16)(16+1)}{36(36+1)} = 180.13$$

Hence, $180.13 was saved.

21. $f(18 \times 13.28) - 200 = 39.04$
$k = 18 - 10 = 8$

$$u = \frac{fk(k+1)}{n(n+1)} = \frac{39.04(8)(8+1)}{18(18+1)} = 8.22$$

Hence, $8.22 was saved.

23. $f = (10 \times 99.75) - 950 = 47.50$
$k = 10 - 7 = 3$

$$u = \frac{fk(k+1)}{n(n+1)} = \frac{47.5(3)(3+1)}{10(10+1)} = 5.18$$

Hence, $5.18 was saved.

25. It's a method of borrowing money, whereby portions of the debt (including charges) are paid back from time to time. In the case of credit-card companies, these payments need not be equal or equally spaced. In the case of lending institutions, the payments typically are constant and once a month.

27. The *unpaid balance method* entails interest (for 1 month) charged only on the balance from the previous month. With the *average daily balance method*, the balances for each day of the month are added, and the sum is divided by the number of days involved. Then interest (for 1 month) is charged on the quotient. (That is, on the average daily balance.)

29. The rule of 78s is one method of computing the money saved by a borrower for paying off early (usually a short-term one), and thus reducing the otherwise total interest. It's a fact that
$$\frac{12}{78}+\frac{11}{78}+\frac{10}{78}+\cdots+\frac{1}{78}=\frac{78}{78}=1.$$ For a 12-month loan, this fact is useful to a lending institution that wants to collect during the first month: $\frac{12}{78}$ of the total interest; second month: $\frac{11}{78}$ of the total interest; etc.

31. monthly payment $= \dfrac{P+I}{T}$

$= \dfrac{P+PRT}{T}$

$= \dfrac{P(1+RT)}{T}$

$100 = \dfrac{P[1+(0.08)(4)]}{48}$

$\dfrac{4800}{1+0.32} = P$

$3636.36 = P$

The principal is $3636.36.

33. $I = PRT = (800)(0.12)(1) = 96$,

monthly interest $= \dfrac{96}{12} = 8$

4-month savings $= 4(8) = 32$

If computed equally over 12 months, $32 is saved after 8 payments.

Rule of 78s: $k = 12 - 8 = 4$

$u = \dfrac{fk(k+1)}{n(n+1)} = \dfrac{96(4)(4+1)}{12(12+1)} = 12.31$

Using the rule of 78s, $12.31 is saved. The borrower saves more money when computed equally over 12 months.

Exercise Set 9-4

1. (a) $0.15 \times \$145,000 = \$21,750$

(b) $\$145,000 - \$21,750 = \$123,250$

(c) $\dfrac{\$123,250}{\$1000} \times 7.70 = \$949.03$

(d) $(\$949.03 \times 12 \times 25) - \$123,250 = \$161,459$

3. (a) $0.40 \times \$200,000 = \$80,000$

(b) $\$200,000 - \$80,000 = \$120,000$

(c) $\dfrac{\$120,000}{\$1000} \times 9.52 = \$1142.40$

(d) $(\$1142.40 \times 12 \times 30) - \$120,000$
$= \$291,264$

5. (a) $0.05 \times \$52,500 = \$2625$

(b) $\$52,500 - \$2625 = \$49,875$

(c) $\dfrac{\$49,875}{\$1000} \times \$10.49 = \$523.19$

(d) $(\$523.19 \times 12 \times 40) - \$49,875$
$= \$201,256.20$

7. (a) $0.30 \times \$200,000 = \$60,000$

(b) $\$200,000 - \$60,000 = \$140,000$

(c) $\dfrac{\$140,000}{\$1000} \times \$10.66 = \$1492.40$

(d) $(\$1492.40 \times 12 \times 20) - \$140,000$
$= \$218,176$

9. Step 1 $\quad I = PRT$
$= 123,250(0.07)\left(\dfrac{1}{12}\right)$
$= 718.96$

Step 2 $\quad 949.03 - 718.96 = 230.07$

Step 3 $\quad 123,250 - 230.07 = 123,019.93$

Step 4 $\quad I = 123,019.93(0.07)\left(\dfrac{1}{12}\right) = 717.62$

Step 5 $\quad 949.03 - 717.62 = 231.41$

Step 6 $\quad 123,019.93 - 231.41 = 122,788.52$

Step 7 $I = 122,788.52(0.07)\left(\dfrac{1}{12}\right) = 716.27$

Step 8 $949.03 - 716.27 = 232.76$

Step 9 $122,788.52 - 232.76 = 122,555.76$

Payment number	Interest	Payment on Principal	Balance of Loan
1	$718.96	$230.07	$123,019.93
2	$717.62	$231.41	$122,788.52
3	$716.27	$232.76	$122,555.76

11. Step 1 $I = PRT$

$$= 140,000(0.115)\left(\dfrac{1}{12}\right)$$

$$= 1341.67$$

Step 2 $1492.40 - 1341.67 = 150.73$

Step 3 $140,000 - 150.73 = 139,849.27$

Step 4 $I = 139,849.27(0.115)\left(\dfrac{1}{12}\right)$

$$= 1340.22$$

Step 5 $1492.40 - 1340.22 = 152.18$

Step 6 $139,849.27 - 152.18 = 139,697.09$

Step 7 $I = (139,697.09)(0.115)\left(\dfrac{1}{12}\right)$

$$= 1338.76$$

Step 8 $1492.40 - 1338.76 = 153.64$

Step 9 $139,697.09 - 153.64 = 139,543.45$

Payment number	Interest	Payment on Principal	Balance of Loan
1	$1341.67	$150.73	$139,849.27
2	$1340.22	$152.18	$139,697.09
3	$1338.76	$153.64	$139,543.45

13. $160,000 \times 0.25 = 40,000$
$40,000 \div 1000 \times 75 = 3000$
The property tax is $3000.

15. $375,000 \times 0.50 = 187,500$
$187,500 \div 1000 \times 36.5 = 6843.75$
The property tax is $6843.75.

17. $110,000 \times 0.40 = 44,000$
$44,000 \div 1000 \times 15 = 660$
The property tax is $660.

19. $275,000 \times 0.80 = 220,000$
$220,000 \div 1000 \times 105 = 23,100$
The property tax is $23,100.

21. In a *fixed-rate* mortgage, the rate of interest stays constant throughout the entire loan. In an *adjustable-rate mortgage*, the rate of interest may decrease and/or increase.

23. Multiply the (constant) monthly payment by the number of such payments. Decrease this sum by the amount borrowed. In short:
total interest = (sum of all payments) minus (loan).

25. The *market value* is the present monetary worth of the home. The *assessed value* is a certain percentage of the market value.

27. 25% down payment option:
$0.25 \times \$80,000 = \$20,000$
$\$80,000 - \$20,000 = \$60,000$
$$\dfrac{\$60,000}{\$1000} \times \$9.00 = \$540$$

10% down payment option:
$0.10 \times \$80,000 = \$8000$
$\$80,000 - \$8000 = \$72,000$
$$\dfrac{\$72,000}{\$1000} \times \$7.70 = \$554.40$$
The better option is 20 years at 9% with 25% down payment.

Exercise Set 9-5

1. $M = 0.55 \times \$2250 = \$1237.50$
$S = \$2250 + \$1237.50 = \$3487.50$

3. $M = 0.10 \times \$80 = \$8$
$C = \$80 - \$8 = \$72$

5. $M = \$85 - \$60 = \$25$
$$R = \dfrac{\$25}{\$60} \approx 0.417 = 41.7\%$$

7. $M = \$200 - \$130 = \$70$
$$R = \dfrac{\$70}{\$200} = 0.35 = 35\%$$

9. $C = \$84 - \$40 = \$44$
$$R = \dfrac{\$40}{\$84} \approx 0.476 = 47.6\%$$

11. $C = \dfrac{\$90}{0.45} = \$200$

$S = \$200 + \$90 = \$290$

13. $M = 2.00 \times \$180 = \$360$
$S = \$180 + \$360 = \$540$

15. $C = \dfrac{\$50}{1.50} = \$33.33$

$S = \$33.33 + \$50 = \$83.33$

17. $M = 0.20 \times \$30 = \$6$
$S = \$30 + \$6 = \$36$

19. $C = \dfrac{\$20}{0.3} = \$66.67$

$S = \$66.67 + \$20 = \$86.67$

21. $M = \$70 - \$50 = \$20$

$R = \dfrac{\$20}{\$50} = 0.40 = 40\%$

23. $M = 0.50 \times \$520 = \$260$
$C - \$520 - \$260 = \$260$

25. $S = \dfrac{\$150}{0.80} = \$187.50$

$C = \$187.50 - \$150 = \$37.50$

27. $M = \$15 - \$8 = \$7$

$R = \dfrac{\$7}{\$15} = 0.467 = 46.7\%$

29. $R = \dfrac{\$20}{\$56.99} \approx 0.35 = 35\%$

31. $M = \$60 - \$36 = \$24$

$R = \dfrac{\$24}{\$60} = 0.40 = 40\%$

33. markdown $= 0.40 \times \$98 = \$39.20$
reduced price $= \$98 - \$39.20 = \$58.80$

35. markdown $= 0.60 \times \$15 = \$9$
reduced price $= \$15 - \$9 = \$6$

37. Oct 1: $S = \$40 + 0.20 \times \$40 = \$48$
Nov 3: $S = \$48 - 0.10 \times \$48 = \$43.20$
Dec 1: $S = \$43.20 + 0.30 \times \$43.20 = \$56.16$
Christmas sales: $S = \$56.16 - 0.25 \times \$56.16$
$\qquad\qquad\qquad = \$42.12$
The final selling price is $42.12.

39. Markup rate on selling price $= \dfrac{0.55}{1 + 0.55}$
$\qquad\qquad\qquad\qquad \approx 0.355$
$\qquad\qquad\qquad\qquad = 35.5\%$

41. Markup rate on cost $= \dfrac{0.65}{1 - 0.65} \approx 1.86 = 186\%$

43. The markup on cost is a certain percentage of the cost the merchant paid for some item. The markup on selling price is a certain percentage of the intended selling price. In both cases, the markup is an amount added to the cost of an item, to give the selling price.

45. Selling price is 100% when it is the base.

47. Let x = original selling price
and y = percent markup bringing item back up to x.

$x - 0.2x + y(x - 0.2x) = x$
$(1 - 0.2)x + y(1 - 0.2)x = x$
$0.8x + y(0.8x) = x$
$0.8x(1 + y) = x$
$1 + y = \dfrac{x}{0.8x}$
$1 + y = \dfrac{1}{0.8}$
$y = \dfrac{1}{0.8} - 1$
$y = 0.25$ or 25%

A 25% markup would bring the item back to the original selling price.

Review Exercises

1. $\dfrac{7}{8} = 0.875$
$0.875 = 87.5\%$

3. $80\% = 0.8$
$80\% = \dfrac{80}{100} = \dfrac{4}{5}$

5. $185\% = 1.85$
$185\% = \dfrac{185}{100} = \dfrac{37}{20}$

7. $5\dfrac{3}{4} = 5.75$
$5.75 = 575\%$

9. $45.5\% = 0.455$
$45.5\% = \dfrac{45.5}{100} = \dfrac{455}{1000} = \dfrac{91}{200}$

11. $P = R \times B$
$P = 0.75 \times 96$
$P = 69.12$

13. $P = R \times B$
$275 = 0.25 \times B$
$1100 = B$

15. 6% of what number is $3.60?
$P = R \times B$
$3.6 = (0.06) \times B$
$60 = B$
The cost of the table is $60.

17. $I = PRT = (4300)(0.09)(6) = 2322$
The simple interest is $2322.

19. $I = PRT$
$262.5 = (875)(0.12)(T)$
$$\frac{262.5}{(875)(0.12)} = \frac{(875)(0.12)(T)}{(875)(0.12)}$$
$2.5 = T$
The time is 2.5 years.

21. $I = PRT$
$104.65 = (230)(R)(6.5)$
$$\frac{104.65}{(230)(6.5)} = \frac{(230)(R)(6.5)}{(230)(6.5)}$$
$0.07 = R$
The rate is 7%.

23. $I = PRT$
$385 = P(0.14)(2)$
$$\frac{385}{(0.14)(2)} = \frac{P(0.14)(2)}{(0.14)(2)}$$
$1375 = P$
The principal is $1375.

25. $MV = P\left(1 + \dfrac{r}{n}\right)^{nt}$
$= 1775\left(1 + \dfrac{0.05}{1}\right)^{1 \cdot 6}$
$= 2378.67$
$I = 2378.67 - 1775 = 603.67$
The interest is $603.67 and the maturity value is $2378.67.

27. $MV = P\left(1 + \dfrac{r}{n}\right)^{nt} = 45\left(1 + \dfrac{0.08}{4}\right)^{4 \cdot 3} = 57.07$

$I = 57.07 - 45 = 12.07$
The interest is $12.07 and the maturity value is $57.07.

29. $I = PRT = 6000(0.06)(5) = 1800$
$MV = 6000 + 1800 = 7800$
The simple interest is $1800 and the maturity value is $7800.

31.
$$I = PRT$$
$$60.48 = (4320)(R)(1)$$
$$\frac{60.48}{(4320)(1)} = \frac{(4320)(R)(1)}{(4320)(1)}$$
$$0.014 = R$$
The rate of interest is 1.4%.

33. $E = \left(1 + \dfrac{r}{n}\right)^n - 1 = \left(1 + \dfrac{0.12}{4}\right)^4 - 1 = 0.1255$

The effective rate is 12.55%.

35. $N = 4 \times 3 = 12;\ R = \dfrac{0.03}{4} = 0.0075$

$$\text{FV} = P\left(\frac{(1+R)^N - 1}{R}\right)$$
$$= 650\left(\frac{(1+0.0075)^{12} - 1}{0.0075}\right)$$
$$= 8129.93$$

The future value of the annuity is $8129.93.

37.

Date	Balance	Days	Balance × days
4-1	5628.00	$10 - 1 = 9$	50,652
4-10	$5628 + 2134.60 = 7762.60$	$22 - 10 = 12$	93,151.20
4-22	$7762.60 - 900 = 6862.60$	$28 - 22 = 6$	41,175.60
4-28	$6862.60 + 437.80 = 7300.40$	$30 - 28 + 1 = 3$	21,901.2
Total		30	206,880

$$\text{average daily balance} = \frac{206,880}{30} = 6896$$

$(6896)(0.018) = 124.13$
The finance charge is $124.13.
$7300.40 + 124.13 = 7424.53$
The new balance is $7424.53.

39. $f = (30 \times 61.25) - 1500 = 337.50$
$k = 30 - 24 = 6$
$$u = \frac{fk(k+1)}{n(n+1)} = \frac{337.5(6)(6+1)}{30(30+1)} \approx 15.24$$
Hence, $15.24 is saved.

41. $135,000 \times 0.30 = 40,500$
$40,500 \div 1000 \times 70 = 2835$
The property tax is $2835.

43. $M = 0.18 \times \$25 = \$4.50$
$C = \$25 - \$4.50 = \$20.50$

45. $M = \$900 - \$675 = \$225$
$$R = \frac{\$225}{\$900} = 0.25 = 25\%$$

47. Markup rate on cost $= \dfrac{0.42}{1 - 0.42} \approx 0.724 = 72.4\%$

Chapter Test

1. $\dfrac{5}{16} = 0.3125 = 31.25\%$

3. $28\% = \dfrac{28}{100} = \dfrac{7}{25}$

5. $P = R \times B$
$32 = R \times 40$
$0.8 = R$
32 is 80% of 40.

7. $P = R \times B$
$135 = 0.45 \times B$
$300 = B$
45% of 300 is 135.

9. $385.20 is 15% of what amount?
$P = R \times B$
$385.2 = 0.15 \times B$
$2568 = B$
The person sold $2568 worth of merchandise.

11. $I = PRT$
$I = (1350)(0.12)(3) = 486$
The simple interest is $486.

13. $I = PRT = (435)(0.0375)(0.5) = 8.16$
$MV = 435 + 8.16 = 443.16$
monthly payment $= \dfrac{443.16}{6} = 73.86$
The interest is $8.16 and the maturity value is $443.16. The monthly payment is $73.86.

15. $I = PRT = (1800)(0.12)(1) = 216$
$MV = 1800 + 216 = 2016$
monthly payment $= \dfrac{2016}{12} = 168$
The interest is $216 and the maturity value is $2016. The monthly payment is $168.

17. $MV = P\left(1 + \dfrac{r}{n}\right)^{nt}$
$= 9750\left(1 + \dfrac{0.1}{4}\right)^{4 \cdot 6}$
$= 17,635.08$
$I = 17,635.08 - 9750 = 7885.08$
The interest is $7885.08 and the maturity value is $17,635.08.

19. $E = \left(1 + \dfrac{r}{n}\right)^{n} - 1 = \left(1 + \dfrac{0.08}{2}\right)^{2} - 1 = 0.0816$
The effective rate is 8.16%.

21. $I = PRT = 1250\,(0.016)(1) = 20$
The finance charge is $20.
New balance $= 1250 + 20 + 560 - 800 = 1030$
The new balance is $1030.

23. $I = (12 \times 70.70) - 800 = 48.40$
$\text{APR} \approx \dfrac{2NI}{P(T+1)} \approx \dfrac{2(12)(48.40)}{800(12+1)} \approx 0.1117$
The annual percentage rate is approximately 11.17%.

25. (a) $0.05 \times \$80,000 = \$4000$

(b) $\$80,000 - \$4000 = \$76,000$

(c) $\dfrac{\$76,000}{\$1000} \times \$8.05 = \$611.80$

(d) Step 1 $\quad I = PRT$
$= (76,000)(0.09)\left(\dfrac{1}{12}\right)$
$= \$570$
Step 2 $\quad 611.80 - 570 = 41.80$
Step 3 $\quad 76,000 - 41.80 = 75,958.20$
Step 4 $\quad I = (75,958.20)(0.09)\left(\dfrac{1}{12}\right)$
$= 569.69$
Step 5 $\quad 611.80 - 569.69 = 42.11$
Step 6 $\quad 75,958.20 - 42.11 = 75,916.09$

Payment number	Interest	Payment on Principal	Balance of Loan
1	$570	$41.80	$75,958.20
2	$569.69	$42.11	$75,916.09

27. $M = 25 - 15 = 10$
Markup rate on cost $= \dfrac{10}{15} \approx 0.667 = 66.7\%$
Markup rate on selling price $= \dfrac{10}{25} = 0.40 = 40\%$

29. Markup rate on cost $= \dfrac{0.20}{1 - 0.20} = 0.25 = 25\%$

1

10 Geometry

Exercise Set 10-1

1. Open segment; $\overset{\circ\!\!-\!\!\circ}{PQ}$

3. Line; $\overleftrightarrow{RS}$

5. Half open segment; $\overset{\bullet\!-\!\circ}{CD}$

7. Segment $\overline{TU}$

9. $\angle RST;\ \angle TSR;\ \angle S;\ \angle 3$

11. The angle measures 180°, so it is a straight angle.

13. The angle's measure is between 90° and 180°, so it is an obtuse angle.

15. $\angle 1$ and $\angle 4$ are vertical angles since they are opposite angles formed by two intersecting lines.

17. $\angle 2$ and $\angle 6$ are corresponding angles since they consists of one exterior and one interior angle on the same side of the transversal that intersects two parallel lines.

19. $\angle 1$ and $\angle 5$ are corresponding angles since they consist of one exterior and one interior angle on the same side of the transversal that intersects two parallel lines.

21. $\angle 1$ and $\angle 8$ are alternate exterior angles since they are formed by the transversal that intersects two parallel lines.

23. $90° - 8° = 82°$

25. $90° - 32° = 58°$

27. $90° - 78° = 12°$

29. $180° - 156° = 24°$

31. $180° - 62° = 118°$

33. $180° - 120° = 60°$

35. $m\angle 2 = 143°$ since the two angles are vertical angles.
$m\angle 1 = 180° - m\angle 2 = 180° - 143° = 37°$
since $\angle 1$ and $\angle 3$ are vertical angles,
$m\angle 3 = 37°$.

37. $m\angle 2 = 90°$ since the two angles are vertical angles.
$m\angle 1 = 180° - 90° = 90°$
$m\angle 3 = m\angle 1 = 90°$

39. $m\angle 1 = 180° - 165° = 15°$
$m\angle 1 = m\angle 3 = m\angle 5 = m\angle 7 = 15°$
$m\angle 2 = m\angle 4 = m\angle 6 = 165°$

41. The hands form a right angle, so the measure of the angle is 90°.

43. The hands form an angle that is $\frac{1}{6}$ of the whole circle of 360°, so the measure of the angle is $\frac{360°}{6} = 60°$.

45. A *point* can be considered as a dot, but theoretically smaller than any conceivable size. A *line* is a set of connected points forming a "taut string" of infinite length, but theoretically with no thickness, or widths, or gaps. A *plane* is a flat surface (theoretically without thickness), extending infinitely in every direction.

47. A half line, plus its end point, constitutes a ray.

49. The sum of the measures equals 90° for two complementary angles; 180° for two supplementary angles.

51. $m\angle 1 + m\angle 2 + 70° = 180°$
$m\angle 1 + m\angle 2 = 110°$
$2m\angle 2 = 110°$
$m\angle 2 = 55°$
Since $\angle 2$ and $\angle 3$ are alternate interior angles.
$m\angle 3 = m\angle 2 = 55°$
$m\angle 4 = 180° - 55° = 125°$

Exercise Set 10-2

1. Isosceles triangle since two sides are of equal length, or acute triangle since all angles are acute.

3. Obtuse triangle since one angle is greater than 90°.

5. Scalene triangle since no two sides are of equal length, or acute triangle since all angles are acute.

7. $m\angle A + m\angle B + m\angle C = 180°$
    $60° + 50° + m\angle C = 180°$
    $110° + m\angle C = 180°$
    $m\angle C = 70°$

9. $m\angle A + m\angle B + m\angle C = 180°$
    $140° + 25° + m\angle C = 180°$
    $165° + m\angle C = 180°$
    $m\angle C = 15°$

11. $c^2 = a^2 + b^2$
    $x^2 = 16^2 + 30^2$
    $= 256 + 900$
    $= 1156$
    $x = \sqrt{1156}$
    $= 34$
    Hence, x is 34 feet.

13. $c^2 = a^2 + b^2$
    $650^2 = 330^2 + x^2$
    $422,500 = 108,900 + x^2$
    $313,600 = x^2$
    $\sqrt{313,600} = x$
    $560 = x$
    Hence, x is 560 centimeters.

15. $c^2 = a^2 + b^2$
    $954^2 = 504^2 + x^2$
    $910,116 = 254,016 + x^2$
    $656,100 = x^2$
    $\sqrt{656,100} = x$
    $810 = x$
    Hence, x is 810 yards.

17. $\dfrac{\text{length of side } AB}{\text{length of side } A'B'} = \dfrac{\text{length of side } AC}{\text{length of side } A'C'}$
    $\dfrac{6}{18} = \dfrac{10}{x}$
    $6x = 180$
    $x = 30$
    Hence, x or the length of side $A'C'$ is 30 inches.

19. $\dfrac{\text{length of side } BC}{\text{length of side } BD} = \dfrac{\text{length of side } AB}{\text{length of side } BE}$
    $\dfrac{24}{12} = \dfrac{34}{x}$
    $24x = 408$
    $x = 17$
    Hence, x or the length of side BE is 17 meters.

21. $\dfrac{\text{length of side } BE}{\text{length of side } CD} = \dfrac{\text{length of side } AE}{\text{length of side } AD}$
    $\dfrac{7}{14} = \dfrac{11}{11+x}$
    $7(11+x) = 14(11)$
    $77 + 7x = 154$
    $7x = 77$
    $x = 11$
    Hence, x is 11 feet.

23. $c^2 = a^2 + b^2$
    $c^2 = 60^2 + 60^2$
    $c^2 = 3600 + 3600$
    $c^2 = 7200$
    $c = \sqrt{7200}$
    $c \approx 84.85$
    The distance from home plate to second base is about 84.85 feet.

25. $\dfrac{\text{length of side } CE}{\text{length of side } AC} = \dfrac{\text{length of side } DE}{\text{length of side } AB}$
    $\dfrac{20}{65} = \dfrac{32}{x}$
    $20x = 2080$
    $x = 104$
    The lake is 104 feet wide.

27. $\dfrac{5.5}{165} = \dfrac{9}{x}$
    $5.5x = 1485$
    $x = 270$
    The height of the tree is 270 feet.

29. $c^2 = a^2 + b^2$
    $c^2 = 8^2 + 10^2$
    $c^2 = 64 + 100$
    $c^2 = 164$
    $c = \sqrt{164}$
    $c \approx 12.8$
    The length of the beam is about 12.8 feet.

31. $c^2 = a^2 + b^2$
    $c^2 = 175^2 + 120^2$
    $c^2 = 30,625 + 14,400$
    $c^2 = 45,025$
    $c = \sqrt{45,025}$
    $c \approx 212.2$
    The plane flew about 212.2 miles diagonally.

33. *Equilateral*: all sides equal; *isosceles*: two sides equal; *scalene*: no two sides are equal.

35. Draw a triangle on a piece of paper and cut off the angles. Place the angles next to each other on a flat surface in such a way that their vertices coincide and two pairs of angles have a common side. The set of angles forms a 180° angle.

37. Similar triangles have the same shape, but not necessarily the same size. More specifically, the corresponding angles of such triangles have equal measure.

39.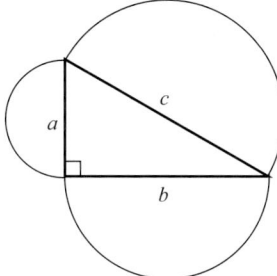

Since the area of a circle is πr^2, the area of a semicircle is $\frac{1}{2}\pi r^2$.

Area of semicircle with diameter a:
$$\frac{1}{2}\pi\left(\frac{a}{2}\right)^2 = \frac{1}{2}\pi\frac{a^2}{4} = \frac{\pi a^2}{8}$$

Area of semicircle with diameter b:
$$\frac{1}{2}\pi\left(\frac{b}{2}\right)^2 = \frac{1}{2}\pi\frac{b^2}{4} = \frac{\pi b^2}{8}$$

Area of semicircle with diameter c:
$$\frac{1}{2}\pi\left(\frac{c}{2}\right)^2 = \frac{1}{2}\pi\frac{c^2}{4} = \frac{\pi c^2}{8}$$

Exercise Set 10-3

1. The polygon has 8 sides, so it is an octagon.
$(n-2)180° = (8-2)180° = 6(180°) = 1080°$
The sum of the measures of the angles is 1080°.

3. The polygon has 3 sides, so it is a triangle.
$(n-2)180° = (3-2)180° = 1(180°) = 180°$
The sum of the measures of the angles is 180°.

5. The polygon has 6 sides, so it is a hexagon.
$(n-2)180° = (6-2)180° = 4(180°) = 720°$
The sum of the measures of the angles is 720°.

7. The quadrilateral is a rectangle since it is a parallelogram with four right angles.

9. The quadrilateral is a trapezoid since it has only two parallel sides.

11. $P = 2l + 2w = 2(22) + 2(16) = 44 + 32 = 76$
The perimeter is 76 yards.

13. $P = 3 + 8 + 10 + 7 = 28$
The perimeter is 28 feet.

15. $P = 6 + 10 + 6 + 6 + 10 + 6 = 44$
The perimeter is 44 inches.

17. The missing sides measure 7 ft, $10 - 5 - 3 = 2$ ft, and 4 ft.
$P = 7 + 5 + 4 + 2 + 4 + 3 + 7 + 10 = 42$
The perimeter is 42 feet.

19. $P = 6s = 6(7) = 42$
The perimeter is 42 miles.

21. $P = 4s = 4(90) = 360$
The player runs at least 360 feet.

23. $P = a + b + c = 62 + 85 + 94 = 241$
241 feet of hedges are needed.

25. With the mat, the dimensions are 15 in. × 18 in. (each side is increased by 2 inches on each end, or 4 inches)
$P = 2l + 2w = 2(15) + 2(18) = 30 + 36 = 66$
66 inches of molding is needed.

27. $P = 2l + 2w = 2(360) + 2(160) = 720 + 320 = 1040$
$\frac{5280}{1040} \approx 5.08$
A person needs to walk around the field about 5.08 times.

29. Similarities:

  (a) In both, opposite sides are $=$ and $\parallel$.

  (b) In both, sum of measures of angles is 360°.

  (c) In both, opposite angles have equal measure.

  Differences:

  (a) In a rhombus, all four sides are $=$ in length; in a rectangle, this need not hold.

  (b) A rectangle always has four right angles; this may, or may not, be true in a rhombus.

31. A square is a rectangle because in both

  (a) Opposite sides are equal in length.

  (b) All four angles are right.

A square is also a rhombus since in both

(a) All four sides are equal in length.

(b) Opposite angles have equal measure.

33. Sum $= (n - 2)180°$, where $n =$ number of sides the polygon has.

35. Since $m\angle ACB + m\angle A + m\angle B = 180°$ and $m\angle ACB + m\angle BCD = 180°$, then $m\angle A + m\angle B = m\angle BCD$.

Exercise Set 10-4

1. $A = s^2 = 17^2 = 289$
The area is 289 square inches.

3. $A = bh = 30(15) = 450$
The area is 450 square yards.

5. $A = bh = (20)(20) = 400$
The area is 400 square meters.

7. $A = \frac{1}{2}bh = \frac{1}{2}(21)(10) = 105$
The area is 105 square miles.

9. $A = \frac{1}{2}h(a+b) = \frac{1}{2}(17)(20+35) = 467.5$
The area is 467.5 square inches.

11. Divide the figure into two rectangles.

$A = lw \qquad\qquad A = lw$
$\quad = 10(6) \qquad\qquad = 4(3)$
$\quad = 60 \qquad\qquad\quad = 12$

The total area is $60 + 12 = 72$ square inches.

13. $C = 2\pi r = 2 \cdot (3.14) \cdot 8 = 50.24$

$A = \pi r^2 = 3.14 \cdot 8^2 = 200.96$

The circumference is 50.24 inches and the area is 200.96 square inches.

15. $r = \dfrac{d}{2} = \dfrac{16}{2} = 8$ m

$C = 2\pi r = 2 \cdot (3.14) \cdot 8 = 50.24$

$A = \pi r^2 = 3.14 \cdot 8^2 = 200.96$

The circumference is 50./24 meters and the area is 200.96 square meters.

17. $C = 2\pi r = 2 \cdot (3.14) \cdot 21 = 131.88$

$A = \pi r^2 = 3.14 \cdot 21^2 = 1384.74$

The circumference is 131.88 kilometers and the area is 1384.74 square kilometers.

19. $A = s^2 = 10^2 = 100$ ft^2

$\dfrac{100}{9} \approx 11.11$ yd^2

11.11 square yards of vinyl linoleum are needed.

21. $\dfrac{25}{5} = 5$ for the length

$\dfrac{24}{2} = 12$ for the width

$5 \times 12 = 60$

60 tickets can be cut from the poster board.

23. $A = \frac{1}{2}h(a+b) = \frac{1}{2}(40)(60+75) = 2700$

$2700 \cdot (\$0.60) = \$1620$

The plywood will cost $1620.

25. $A = \frac{1}{2}bh = \frac{1}{2}(5)(3) = 7.5$

The area of the shelf is 7.5 square feet.

27. $C = \pi r^2 = \pi \left(\dfrac{2}{12}\right)^2 = 3.14(6)^2 = 113.04$ ft^2

$\dfrac{113.04}{9} = 12.56$ yd^2

$12.56 \times \$15 = \$188.40$

The carpeting will cost $188.40.

29. $A = \pi r^2 = 3.14 \cdot 60^2 = 11,304$
The broadcast area is 11,304 square miles.

31. $A = \frac{1}{2}\pi r^2 = \frac{1}{2}(3.14)\left(\dfrac{4}{2}\right)^2 = 6.28$

$A = \frac{1}{2}bh = \frac{1}{2}(4)(5) = 10$

$6.28 + 10 = 16.28$

The area of the material is 16.28 square feet.

33. The *perimeter* is the distance around the border of the polygon. The *area* is the number of unit squares needed to cover completely the space enclosed by the polygon, but no other space. (A "unit square" means a square, each of whose sides measures 1 cm, or 1 inch, or 1 ft, etc.)

35. Take any circle; the circumference divided by the diameter is always the same answer, which we denote by π.

37. $s = \dfrac{1}{2}(a+b+c) = \dfrac{1}{2}(5+12+13) = \dfrac{1}{2}(30) = 15$

$$A = \sqrt{s(s-a)(s-b)(s-c)}$$
$$= \sqrt{15(15-5)(15-12)(15-13)}$$
$$= \sqrt{15(10)(3)(2)}$$
$$= \sqrt{900}$$
$$= 30$$

The area is 30 square inches.

Exercise Set 10-5

1. $SA = 6s^2 = 6(5^2) = 150$

$V = s^3 = 5^3 = 125$

The surface area is 150 square inches and the volume is 125 cubic inches.

3. $SA = 2lw + 2lh + 2wh$
$$= 2(7)(6) + 2(7)(5) + 2(6)(5)$$
$$= 84 + 70 + 60$$
$$= 214$$
$V = lwh = (7)(6)(5) = 210$

The surface area is 214 square meters and the volume is 210 cubic meters.

5. First, find the slant height, s of the triangular face.

$s = \sqrt{4^2 + 3^2} = \sqrt{25} = 5$

Area of triangular face $= \dfrac{1}{2}bh$ or $\dfrac{1}{2}bs$
$$= \dfrac{1}{2}(6)(5)$$
$$= 15$$

Area of square base, $B = s^2 = 6^2 = 36$

$SA = 4(15) + 36 = 96$

$V = \dfrac{1}{3}Bh = \dfrac{1}{3}(36)(4) = 48$

The surface area is 96 square meters and the volume is 48 cubic meters.

7. $r = \dfrac{d}{2} = \dfrac{28}{2} = 14$

$SA = 2\pi r^2 + 2\pi rh$
$$= 2(3.14)(14)^2 + 2(3.14)(14)(32)$$
$$= 4044.32$$

$V = \pi r^2 h = 3.14(14)^2(32) = 19,694.08$

The surface area is 4044.32 square centimeters and the volume is 19,694.08 cubic centimeters.

9. $s = \sqrt{r^2 + h^2} = \sqrt{8^2 + 15^2} = \sqrt{289} = 17$

$SA = \pi r^2 + \pi rs$
$$= (3.14)(8^2) + (3.14)(8)(17)$$
$$= 628$$

$V = \dfrac{1}{3}\pi r^2 h = \dfrac{1}{3}(3.14)(8^2)(15) = 1004.8$

The surface area is 628 square feet and the volume is 1004.8 cubic feet.

11. $SA = 4\pi r^2 = 4(3.14)(20^2) = 5024$

$V = \dfrac{4}{3}\pi r^3 = \dfrac{4}{3}(3.14)(20^3) \approx 33,493.33$

The surface area is 5024 square inches and the volume is 33,493.33 cubic inches.

13. Subtract the volume of the inner cylinder from that of the outer cylinder.

$V = \pi r_1^2 h - \pi r_2^2 h$
$$= (3.14)\left(\dfrac{9}{2}\right)^2(10.5) - (3.14)\left(\dfrac{5}{2}\right)^2(10.5)$$
$$= 461.58$$

The volume of the shaded area is 461.58 cubic centimeters.

15. $V = lwh = (18)(12)(3) = 648$

648 cubic feet of dirt must be removed.

17. $B = s^2 = 932^2 = 868,624$

$V = \dfrac{1}{3}Bh = \dfrac{1}{3}(868,624)(657) = 190,228,656$

The volume of the pyramid is 190,228,656 cubic feet.

19. $r = \dfrac{d}{2} = \dfrac{22}{2} = 11$

$V = \pi r^2 h = (3.14)(11^2)(36) = 13,677.84$

The volume of the tank is 13,677.84 cubic inches.

21. $s = \sqrt{(3.5)^2 + (11)^2} = \sqrt{133.25}$

The surface area does not include the base.

$SA = \pi rs = (3.14)(3.5)\left(\sqrt{133.25}\right) \approx 126.86$

The surface area of the cone is 126.86 square inches.

23. $r = \dfrac{d}{2} = \dfrac{4200}{2} = 2100$

$V = \dfrac{4}{3}\pi r^3 = \dfrac{4}{3}(3.14)(2100)^3 \approx 3.88 \times 10^{10}$

The volume is 3.88×10^{10} cubic miles.

25. Pretend we can detach a very thin skin off the outermost surface of the solid. Lay this skin flat on a flat surface and measure the area covered.

27. Subtract the volume of all the balls from that of the box.

$$\text{Volume of one ball} = \frac{4}{3}\pi r^3$$

$$= \frac{4}{3}(3.14)\left(\frac{3}{2}\right)^3$$

$$= 14.13$$

Volume of box $= lwh = 12 \cdot 9 \cdot 3 = 324$

Volume left over $= 324 - 12(14.13) = 154.44$

The volume of the space left over is approximately 154 cubic inches.

Exercise Set 10-6

1.

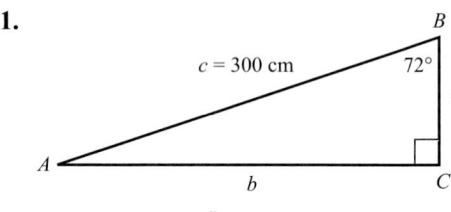

$$\cos B = \frac{a}{c}$$

$$\cos 72° = \frac{a}{300}$$

$$0.3090 = \frac{a}{300}$$

$$300(0.3090) = a$$

$$92.7 = a$$

$$a = 92.7 \text{ cm}$$

3.

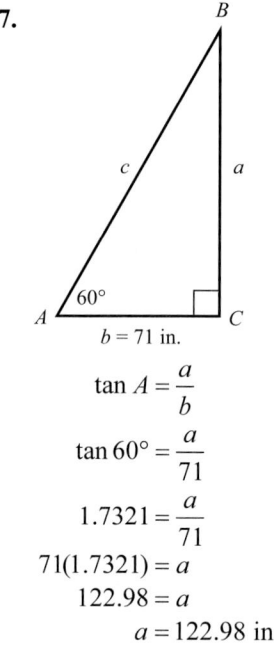

$$\cos A = \frac{b}{c}$$

$$\cos 76° = \frac{18.6}{c}$$

$$0.2419 = \frac{18.6}{c}$$

$$c = \frac{18.6}{0.2419}$$

$$c = 76.89 \text{ in.}$$

5.

$$\sin B = \frac{b}{c}$$

$$\sin 8° = \frac{b}{10}$$

$$0.1392 = \frac{b}{10}$$

$$10(0.1392) = b$$

$$1.392 = b$$

$$b = 1.392 \text{ mm}$$

7.

$$\tan A = \frac{a}{b}$$

$$\tan 60° = \frac{a}{71}$$

$$1.7321 = \frac{a}{71}$$

$$71(1.7321) = a$$

$$122.98 = a$$

$$a = 122.98 \text{ in.}$$

9.

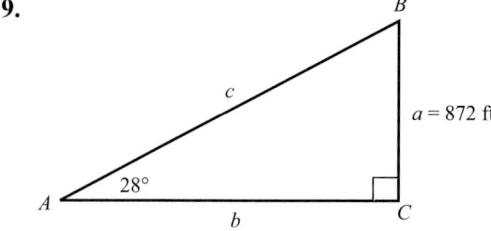

$$\sin A = \frac{a}{c}$$

$$\sin 28° = \frac{872}{c}$$

$$0.4695 = \frac{872}{c}$$

$$c = \frac{872}{0.4695}$$

$$c = 1857.29 \text{ ft}$$

11.

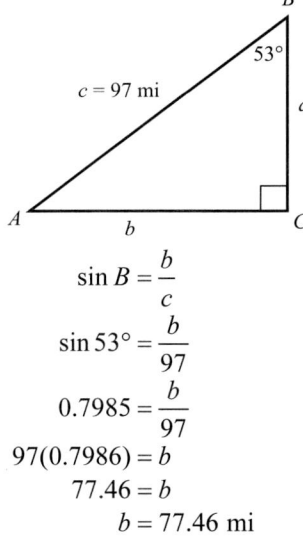

$$\sin B = \frac{b}{c}$$

$$\sin 53° = \frac{b}{97}$$

$$0.7985 = \frac{b}{97}$$

$$97(0.7986) = b$$

$$77.46 = b$$

$$b = 77.46 \text{ mi}$$

13.

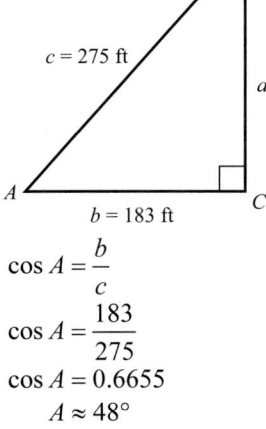

$$\cos A = \frac{b}{c}$$

$$\cos A = \frac{183}{275}$$

$$\cos A = 0.6655$$

$$A \approx 48°$$

15.

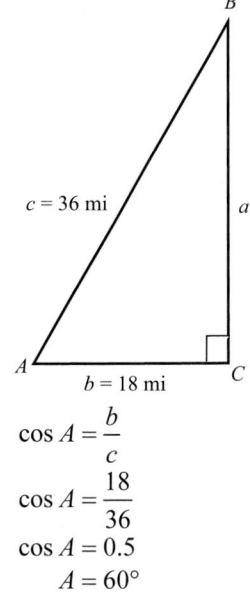

$$\cos A = \frac{b}{c}$$

$$\cos A = \frac{18}{36}$$

$$\cos A = 0.5$$

$$A = 60°$$

17.

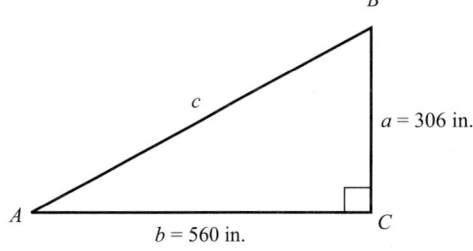

$$\tan A = \frac{a}{b}$$

$$\tan A = \frac{306}{560}$$

$$\tan A = 0.5464$$

$$A \approx 29°$$

19.

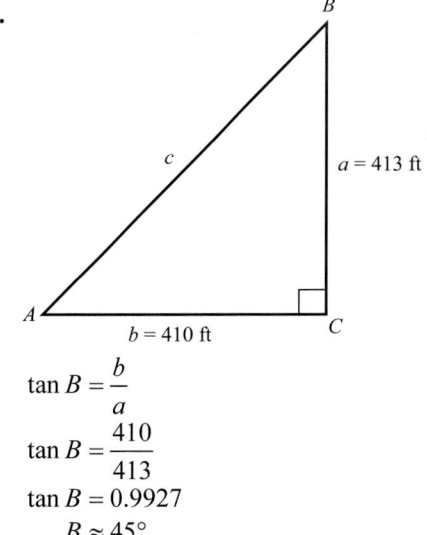

$$\tan B = \frac{b}{a}$$

$$\tan B = \frac{410}{413}$$

$$\tan B = 0.9927$$

$$B \approx 45°$$

21.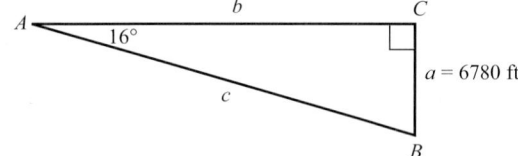

Find b.

$$\tan A = \frac{a}{b}$$

$$\tan 16° = \frac{6780}{b}$$

$$0.2867 = \frac{6780}{b}$$

$$b = \frac{6780}{0.2867}$$

$$b = 23,648.41$$

The plane is about 23,648.41 feet from the tower.

23.

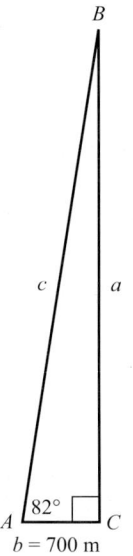

Find a.

$$\tan A = \frac{a}{b}$$

$$\tan 82° = \frac{a}{700}$$

$$7.1154 = \frac{a}{700}$$

$$700(7.1154) = a$$

$$4980.78 = a$$

The height of the weather balloon is 4980.78 meters.

25. Find a.

$$\tan A = \frac{a}{b}$$

$$\tan 25° = \frac{a}{2614}$$

$$0.4663 = \frac{a}{2614}$$

$$2614(0.4663) = a$$

$$1218.91 = a$$

The length of the lake is 1218.91 feet.

27.

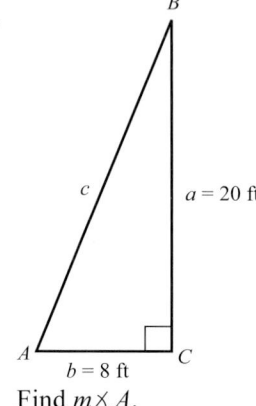

Find $m\angle A$.

$$\tan A = \frac{a}{b}$$

$$\tan A = \frac{20}{8}$$

$$\tan A = 2.5$$

$$A \approx 68°$$

The angle of elevation of the sun is about 68°.

29.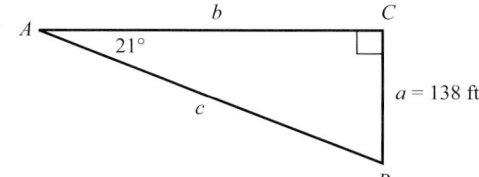

Find b.

$$\tan A = \frac{a}{b}$$

$$\tan 21° = \frac{138}{b}$$

$$0.3839 = \frac{138}{b}$$

$$b = \frac{138}{0.3839}$$

$$b = 359.47$$

The ship is 359.47 feet from the base of the lighthouse.

31.

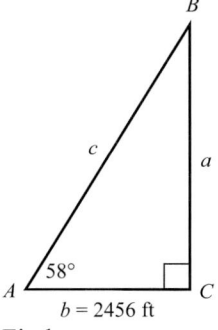

Find a.

$$\tan A = \frac{a}{b}$$

$$\tan 58° = \frac{a}{2456}$$

$$1.6003 = \frac{a}{2456}$$

$$2456(1.6003) = a$$

$$3930.34 = a$$

The height of the cloud is 3930.34 feet.

33. Triangle measurement.

35. All right triangles having a 30° angle are similar. Hence, given any one of these triangles, the ratio of any two sides equals the corresponding ratio from any other of said triangles.

37. It's the angle made by a horizontal line and the downward line of sight to an object.

39.

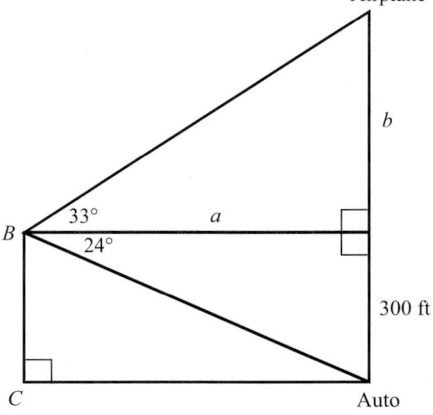

Airplane

a = distance from auto to base of building

$b + 300$ = height of plane

$$\tan 24° = \frac{300}{a}$$

$$0.4452 = \frac{300}{a}$$

$$a = \frac{300}{0.4452}$$

$$a = 673.8$$

$$\tan 33° = \frac{b}{673.8}$$

$$0.6494 = \frac{b}{673.8}$$

$$673.8(0.6494) = b$$

$$437.6 = b$$

$$437.6 + 300 = 737.6$$

The height of the plane is 737.6 feet and the distance the automobile is from the base of the building is 673.8 feet.

Exercise Set 10-7

1. There are 5 evens and 4 odds. The network is not traversable since property 4 states that if a network has more than two odd vertices, it is impossible to traverse.

3. There are 2 evens and 2 odds. The network is traversable provided that one starts at one of the odd vertices and ends at the other odd vertex (property 3).

5. There are 3 evens and 2 odds. The network is traversable provided that one starts at one of the odd vertices and ends at the other odd vertex (property 3).

7. There are 3 evens and 2 odds. The network is traversable provided that one starts at one of the odd vertices and ends at the other odd vertex (property 3).

9. There are 3 evens and 4 odds. The network is not traversable since property 4 states that if a network has more than two odd vertices, it is impossible to traverse.

11.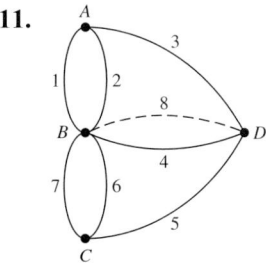

Solution I: Add bridge 8. Then A = odd, D = even; B = even; C = odd; hence, network is traversable.
Solution II: Do not have bridge 8 or bridge 2. Then A = even, B = even, C = odd, D = odd; hence, network is traversable. *Note:* Other solutions do exist.

13. A network is traversable if it's possible to pass through, or trace, each path exactly once, without lifting your pencil. However, a vertex can be crossed more than once.

15.

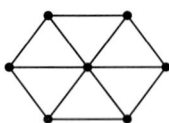

One possibility: The vertex at the center is even; all other vertices are odd.

17. Answers will vary.

Review Exercises

1. The figure is a line.

3. The figure is an open segment.

5. The figure is a trapezoid since it is a quadrilateral with only two parallel sides.

7. The figure is a triangle since it is a closed geometric figure consisting of three sides and three angles. It is an equilateral triangle since it has three sides of equal length.

9. The figure is a hexagon since it is a closed geometric figure consisting of six sides. It is a regular hexagon since it has six sides of equal length.

11. The angle is obtuse since its measure is between $90°$ and $180°$.

13. The angles are vertical angles since they are opposite angles formed by two intersecting lines.

15. The angles are alternate exterior angles since they are the opposite exterior angles formed by the transversal that intersects two parallel lines.

17. (a) $90° - 27° = 63°$

(b) $90° - 88° = 2$

19. $180° - (95° + 42°) = 180° - 137° = 43°$

21. $\dfrac{\text{height of tower}}{\text{height of pole}} = \dfrac{\text{shadow of tower}}{\text{shadow of pole}}$

$$\frac{x}{6} = \frac{24}{2}$$
$$2x = 6(24)$$
$$2x = 144$$
$$x = 72$$

The height of the tower is 72 feet.

23. $m \angle 5 = 180° - 42° = 138°$
$m \angle 1 = m \angle 4 = m \angle 5 = m \angle 7 = 138°$
$m \angle 2 = m \angle 3 = m \angle 6 = 42°$

25. $P = 4 + 3 + 3 + 3 + 3 = 16$
The perimeter is 16 feet.

27. $A = \dfrac{1}{2}bh = \dfrac{1}{2}(16)(7.6) = 60.8$
The area is 60.8 square meters.

29. $A = \dfrac{1}{2}h(a+b) = \dfrac{1}{2}(8)(12+17) = 116$
The area is 116 square feet.

31. $V = \dfrac{4}{3}\pi r^3 = \dfrac{4}{3}(3.14)(18)^3 = 24{,}416.64$
The volume is 24,416.64 cubic centimeters.

33. $V = \dfrac{1}{3}\pi r^2 h = \dfrac{1}{3}(3.14)(3^2)(9) = 84.78$
The volume is 84.78 cubic yards.

35. $SA = 4\pi r^2 = 4(3.14)(5.2)^2 = 339.62$
The surface area is 339.62 square inches.

37. Find the circumference.
$C = \pi d = 3.14(26) = 81.64$ in.
$5280 \text{ ft} \times \dfrac{12 \text{ in.}}{1 \text{ ft}} = 63{,}360$ in.
$\dfrac{63{,}360}{81.64} \approx 776.09$
The wheel will make 776.09 revolutions.

39. $r = \dfrac{d}{2} = \dfrac{4.2}{2} = 2.1$

$V = \dfrac{4}{3}\pi r^3 = \dfrac{4}{3}(3.14)(2.1)^3 \approx 38.77$
The volume is 38.77 cubic centimeters.

41.

Find $m \angle A$.
$\tan A = \dfrac{a}{b}$
$\tan A = \dfrac{32}{40}$
$\tan A = 0.8$
$A \approx 39°$
The angle of elevation of the sun is about $39°$.

43. The network has 2 odd vertices and 4 even vertices. The network is traversable provided that one starts at one of the odd vertices and ends at the other odd vertex (property 3).

45. The network has 4 odd vertices and 1 even vertex. The network is not traversable since property 4 states that if a network has more than two odd vertices, it is impossible to traverse.

47. The network has 5 even vertices and 2 odd vertices. The network is traversable provided that one starts at one of the odd vertices and ends at the other odd vertex (property 3).

Chapter Test

1. $90° - 73° = 17°$

3. $m\angle 2 = 28°$ since the two angles are vertical angles.
$m\angle 1 = 180° - 28° = 152°$
$m\angle 3 = m\angle 1 = 152°$

5. $180° - (85° + 47°) = 180° - 132° = 48°$
The third angle measures $48°$.

7. $\dfrac{\text{length of side } AC}{\text{length of side } A'C'} = \dfrac{\text{length of side } BC}{\text{length of side } B'C'}$
$$\frac{8}{2} = \frac{10}{x}$$
$$8x = 20$$
$$x = 2.5$$
Hence, x or the length of side $B'C'$ is 25 feet.

9. An octagon has 8 sides.
$(n - 2)180° = (8 - 2)180° = 6(180°) = 1080°$
The sum of the angles is $1080°$.

11. $A = lw = (7.5)(2.4) = 18$
The area is 18 square inches.

13. $A = \dfrac{1}{2}h(a + b) = \dfrac{1}{2}(18)(20 + 28) = 432$
The area is 432 square yards.

15. $A = s^2 = (18.6)^2 = 345.96$
The area is 345.96 square miles.

17. $V = s^3 = (9.6)^3 \approx 884.74$
The volume is 884.74 cubic inches.

19. $V = \dfrac{1}{3}Bh = \dfrac{1}{3}(5^2)(12) = 100$
The volume is 100 cubic feet.

21. $V = lwh = (10)(5)(3) = 150$
The volume is 150 cubic feet.

23. $s = \sqrt{20^2 + 12^2}$
$s = \sqrt{544}$
$SA = \pi r^2 + \pi rs$
$ = (3.14)(12)^2 + (3.14)(12)\left(\sqrt{544}\right)$
$ \approx 1331.00$
The surface area is 1331 square feet.

25. $SA = 4\pi r^2 = 4(3.14)\left(5\dfrac{1}{4}\right)^2 \approx 346.2$
The surface area is 346.2 square inches.

27. $c^2 = a^2 + b^2$
$25^2 = 5^2 + b^2$
$625 = 25 + b^2$
$600 = b^2$
$\sqrt{600} = b$
$24.49 \approx b$
The ladder reaches 24.49 feet up the wall.

29. $V = \dfrac{1}{3}\pi r^2 h = \dfrac{1}{3}(3.14)\left(\dfrac{1}{2}\right)^2(2) \approx 0.52$

$0.52 \times 0.41 \approx 0.21$
The sinker weighs 0.21 pound.

31.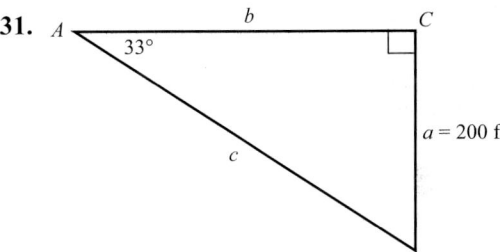

The top of the lighthouse is at A and the boat is at B. Find b.

$\tan A = \dfrac{a}{b}$
$\tan 33° = \dfrac{200}{b}$
$0.6494 = \dfrac{200}{b}$
$b = \dfrac{200}{0.6494}$
$b \approx 307.98$
The boat is 307.98 feet from the lighthouse.

33. There are 8 even vertices and 0 odd vertices. The network is traversable since property 2 states that if all the vertices are even, it can be traversed.

35. There are 4 odd vertices and 3 even vertices. The network is not traversable since property 4 states that if a network has more than two odd vertices, it is impossible to traverse.

11 Probability and Counting Techniques

Exercise Set 11-1

1. The sample space is {1, 2, 3, 4, 5, 6}. Since there are six outcomes, $n(S) = 6$.

(a) $P(4) = \dfrac{n(E)}{n(S)} = \dfrac{1}{6}$

(b) $P(\text{even number}) = \dfrac{n(E)}{n(S)} = \dfrac{3}{6} = \dfrac{1}{2}$

(c) $P(\text{a number greater than 4}) = \dfrac{n(E)}{n(S)} = \dfrac{2}{6} = \dfrac{1}{2}$

(d) $P(\text{a number less than 7}) = \dfrac{n(E)}{n(S)} = \dfrac{6}{6} = 1$

(e) $P(\text{a number greater than 0}) = \dfrac{n(E)}{n(S)} = \dfrac{6}{6} = 1$

(f) $P(\text{a number greater than 3 or an odd number})$
$= \dfrac{n(E)}{n(S)}$
$= \dfrac{5}{6}$

(g) $P(\text{a number greater than 3 and an odd number})$
$= \dfrac{n(E)}{n(S)}$
$= \dfrac{1}{6}$

3. The sample space is {1, 2, 3, 4, 5, 6, 7}. Since there are seven outcomes $n(S) = 7$.

(a) $P(6) = \dfrac{n(E)}{n(S)} = \dfrac{1}{7}$

(b) $P(\text{an even number}) = \dfrac{n(E)}{n(S)} = \dfrac{3}{7}$

(c) $P(\text{a number greater than 4}) = \dfrac{n(E)}{n(S)} = \dfrac{3}{7}$

(d) $P(\text{a number less than 8}) = \dfrac{n(E)}{n(S)} = \dfrac{7}{7} = 1$

(e) $P(\text{a number greater than 7}) = \dfrac{n(E)}{n(S)} = \dfrac{0}{7} = 0$

5. $P(\text{woman}) = \dfrac{f}{n} = \dfrac{5}{9}$

7. $P(\text{man}) = \dfrac{f}{n} = \dfrac{9}{16}$

9. $P(\text{captured}) = \dfrac{f}{n} = \dfrac{12,166}{12,249} \approx 0.99$

11. $P(\text{does not use flattery}) = 1 - P(\text{uses flattery})$
$= 1 - 0.16$
$= 0.84$
$= 84\%$

13. The sample space is
{1, 2, 3, 4, 5, 6, 7, 8, 9, 10, 11, 12, 13, 14, 15}.
Since there are 15 outcomes, $n(S) = 15$.

(a) $P(\text{odd number}) = \dfrac{n(E)}{n(S)} = \dfrac{8}{15}$

(b) $P(\text{divisible by 3}) = \dfrac{n(E)}{n(S)} = \dfrac{5}{15} = \dfrac{1}{3}$

(c) $P(\text{multiple of 5}) = \dfrac{n(E)}{n(S)} = \dfrac{3}{15} = \dfrac{1}{5}$

(d) $P(\text{greater than 10}) = \dfrac{n(E)}{n(S)} = \dfrac{5}{15} = \dfrac{1}{3}$

(e) $P(\text{less than 4}) = \dfrac{n(E)}{n(S)} = \dfrac{3}{15} = \dfrac{1}{5}$

15. The sample space is {$1, $5, $10, $20, $50}. Since there are five outcomes, $n(S) = 5$.

(a) $P(\$20) = \dfrac{n(E)}{n(S)} = \dfrac{1}{5}$

(b) $P(\text{larger than } \$5) = \dfrac{n(E)}{n(S)} = \dfrac{3}{5}$

(c) $P(\text{contains the digit 5}) = \dfrac{n(E)}{n(S)} = \dfrac{2}{5}$

(d) $P(\text{does not contain 0}) = \dfrac{n(E)}{n(S)} = \dfrac{2}{5}$

17. A probability experiment is a process that leads to well-defined results called outcomes.

19. An event is a subset whose elements are outcomes.

21. From zero (inclusive) to one (inclusive)

23. 0

25. The sample space consists of two outcomes: rain and no rain. Since all the probabilities add to 1, we have $P(\text{rain}) + P(\text{no rain}) = 1$, whence $0.45 + P(\text{no rain}) = 1$, $P(\text{no rain}) = 0.55$.

27. The probability of an event will always be a number between and including zero and one. In other words, probabilities cannot be negative or greater than one. (b), (d), (f), and (i) cannot be considered a probability of an event.

29. Answers will vary.

31. There are six outcomes possible for each roll. For three rolls, there are $6 \cdot 6 \cdot 6 = 216$ possible outcomes. there are six possible triples, namely, 111, 222, 333, 444, 555, or 666.

$$P(\text{triple}) = \frac{n(E)}{n(S)} = \frac{6}{216} = \frac{1}{36}$$

Exercise Set 11-2

1. Since there are eight outcomes, $n(S) = 8$.

(a) $n(E) = 3$; namely, BGG, GBG, or GGB.

$$P(\text{two girls}) = \frac{n(E)}{n(S)} = \frac{3}{8}$$

(b) $n(E) = 1$; namely, BBB.

$$P(\text{three boys}) = \frac{n(E)}{n(S)} = \frac{1}{8}$$

(c) $n(E) = 7$; namely, BBG, BGB, BGG, GBB, GBG, GGB, or GGG.

$$P(\text{at least one girl}) = \frac{n(E)}{n(S)} = \frac{7}{8} \text{ or}$$

$$P(\text{at least one girl}) = 1 - P(\text{three boys})$$
$$= 1 - \frac{7}{8}$$
$$= \frac{7}{8}$$

3. Since there are 12 outcomes, $n(S) = 12$.

(a) $n(E) = 5$; namely, T2, T3, T4, T5, or T6. $P(\text{tail on coin and number greater than 1 on die})$

$$= \frac{n(E)}{n(S)}$$
$$= \frac{5}{12}$$

(b) $n(E) = 3$; namely H2, H4, or H6.

$$P(\text{head on coin and even on die}) = \frac{n(E)}{n(S)}$$
$$= \frac{3}{12}$$
$$= \frac{1}{4}$$

(c) $n(E) = 2$; namely, T3 or T6. $P(\text{tail on coin and number divisible by 3 on die})$

$$= \frac{n(E)}{n(S)}$$
$$= \frac{2}{12}$$
$$= \frac{1}{6}$$

5. There are three possibilities for the first selection and three for the second. Since there are nine outcomes, $n(S) = 9$.

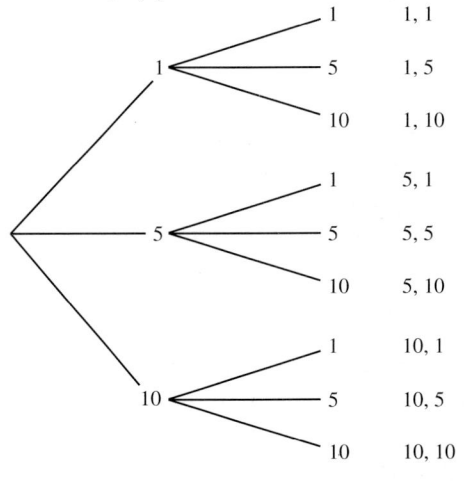

1	1, 1
5	1, 5
10	1, 10
1	5, 1
5	5, 5
10	5, 10
1	10, 1
5	10, 5
10	10, 10

(a) $n(E) = 3$; namely: 1, 1; 5, 5; or 10, 10.

$$P(\text{both same value}) = \frac{n(E)}{n(S)} = \frac{3}{9} = \frac{1}{3}$$

(b) $n(E) = 3$; namely: 1, 5; 1, 10; or 5, 10.

$$P(\text{second larger than first}) = \frac{n(E)}{n(S)} = \frac{3}{9} = \frac{1}{3}$$

(c) $n(E) = 1$; namely: 10, 10.

$$P(\text{value of each is even}) = \frac{n(E)}{n(S)} = \frac{1}{9}$$

(d) $n(E) = 4$; namely: 1, 10; 5, 10; 10, 1; or 10, 5.

$$P(\text{value of exactly one is odd}) = \frac{n(E)}{n(S)} = \frac{4}{9}$$

(e) $n(E) = 3$; namely: 1, 1; 1, 5; or 5, 1.

$$P(\text{sum is less than \$10}) = \frac{n(E)}{n(S)} = \frac{3}{9} = \frac{1}{3}$$

7. There are two possibilities for the first game, two for the second game, and two for the third game. Since there are eight outcomes, $n(S) = 8$.

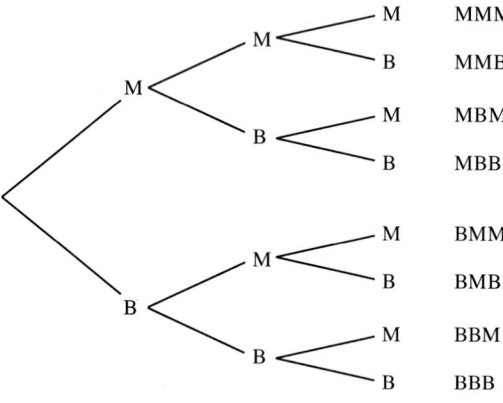

M	M	M	MMM
		B	MMB
	B	M	MBM
		B	MBB
B	M	M	BMM
		B	BMB
	B	M	BBM
		B	BBB

(a) $n(E) = 2$; namely, MMM or BBB.

$$P(\text{either Mark or Bill win all three}) = \frac{n(E)}{n(S)}$$
$$= \frac{2}{8}$$
$$= \frac{1}{4}$$

(b) $n(E) = 6$; namely, MMB, MBM, MBB, BMM, BMB, or BBM.
$P(\text{either Mark or Bill win two out of three})$ or

$$= \frac{n(E)}{n(S)}$$
$$= \frac{6}{8}$$
$$= \frac{3}{4}$$

$P(\text{either Mark or Bill win two out of three})$
$= 1 - P(\text{either Mark or Bill win all three})$
$$= 1 - \frac{1}{4}$$
$$= \frac{3}{4} \text{ (from part a.)}$$

(c) $n(E) = 2$; namely, MMB or BMM.

$$P(\text{Mark wins only two games in a row}) = \frac{n(E)}{n(S)}$$
$$= \frac{2}{8}$$
$$= \frac{1}{4}$$

(d) $n(E) = 1$; namely, BMB.

$$P(\text{Bill wins first, loses second, and wins third}) = \frac{n(E)}{n(S)}$$
$$= \frac{1}{8}$$

9. There are five possibilities for the first selection and four possibilities for the second selection. Since there are 20 outcomes, $n(S) = 20$.

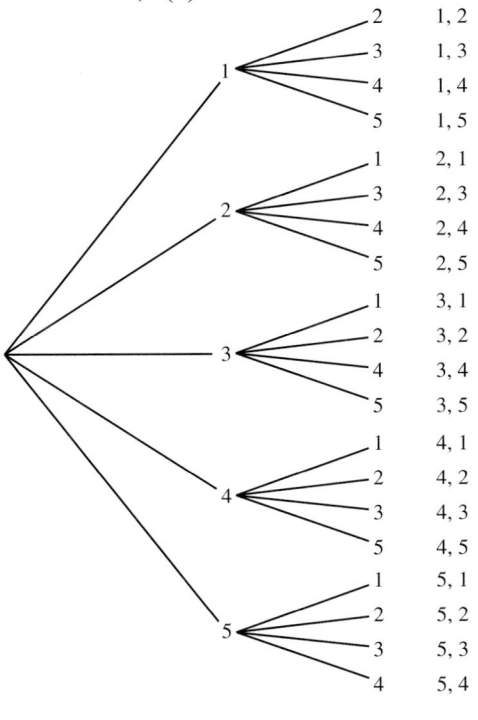

(a) $n(E) = 12$; namely: 1, 2; 1, 4; 2, 1; 2, 3; 2, 5; 3, 2; 3, 4; 4, 1; 4, 3; 4, 5; 5, 2; or 5, 4.

$$P(\text{sum is odd}) = \frac{n(E)}{n(S)} = \frac{12}{20} = \frac{3}{5}$$

(b) $n(E) = 10$; namely: 1, 2; 1, 3; 1, 4; 1, 5; 2, 3; 2, 4; 2, 5; 3, 4; 3, 5; or 4, 5.

$$P(\text{second number larger than first}) = \frac{n(E)}{n(S)}$$
$$= \frac{10}{20}$$
$$= \frac{1}{2}$$

(c) $n(E) = 16$; namely; 1, 4; 1, 5; 2, 3; 2, 4; 2, 5; 3, 2; 3, 4; 3, 5; 4, 1; 4, 2; 4, 3; 4, 5; 5, 1; 5, 2; 5, 3; or 5, 4.

$$P(\text{sum greater than 4}) = \frac{n(E)}{n(S)} = \frac{16}{20} = \frac{4}{5}$$

11. There are three possibilities for MB, two possibilities for monitor, and two possibilities for color. Since there are 12 outcomes, $n(S) = 12$.

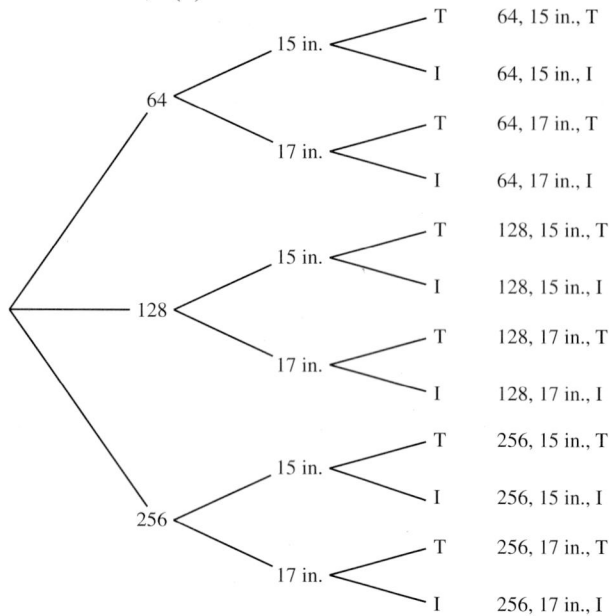

T   64, 15 in., T
I   64, 15 in., I
T   64, 17 in., T
I   64, 17 in., I
T   128, 15 in., T
I   128, 15 in., I
T   128, 17 in., T
I   128, 17 in., I
T   256, 15 in., T
I   256, 15 in., I
T   256, 17 in., T
I   256, 17 in., I

(a) $n(E) = 6$; namely: 64, 17 in., T; 64, 17 in., I; 128, 17 in., T; 128, 17 in., I; 256, 17 in., T; or 256, 17 in., I.

$$P(\text{17-in. monitor}) = \frac{n(E)}{n(S)} = \frac{6}{12} = \frac{1}{2}$$

(b) $n(E) = 2$; namely: 128, 15in., T or 128, 15in., I.

$$P(\text{128 MB and 15 in. monitor}) = \frac{n(E)}{n(S)}$$
$$= \frac{2}{12}$$
$$= \frac{1}{6}$$

(c) $n(E) = 3$; namely: 64, 15 in., I; 128, 15 in., I; or 256, 15 in., I.

$$P(\text{15-in. Ivory monitor}) = \frac{n(E)}{n(S)} = \frac{3}{12} = \frac{1}{4}$$

13. (a) There are four 10s.

$$P(10) = \frac{4}{52} = \frac{1}{13}$$

(b) There are 13 clubs.

$$P(\text{club}) = \frac{13}{52} = \frac{1}{4}$$

(c) There is one ace of hearts.

$$P(\text{ace of hearts}) = \frac{1}{52}$$

(d) There is one 3 and one 5 in each of four suits, so eight possibilities.

$$P(\text{a 3 or a 5}) = \frac{8}{52} = \frac{2}{13}$$

(e) There are four 6s and 13 spades, but the 6 of spades is counted twice in this listing. Hence, there are 16 possibilities.

$$P(6 \text{ or spade}) = \frac{16}{52} = \frac{4}{13}$$

(f) There are four queens and 13 clubs, but the queen of clubs is counted twice in this listing. Hence, there are 16 possibilities.

$$P(\text{queen or club}) = \frac{16}{52} = \frac{4}{13}$$

(g) There are 13 diamonds and 13 clubs, so 26 possibilities.

$$P(\text{diamond or club}) = \frac{26}{52} = \frac{1}{2}$$

(h) There are two red kings.

$$P(\text{red king}) = \frac{2}{52} = \frac{1}{26}$$

(i) There are 26 black cards and four 8s, but the 8 of spades and the 8 of clubs are counted twice in this listing. Hence, there are 28 possibilities.

$$P(\text{black or 8}) = \frac{28}{52} = \frac{7}{13}$$

(j) There are two red 10s.

$$P(\text{red 10}) = \frac{2}{52} = \frac{1}{26}$$

15. There are 36 possible outcomes, so $n(S) = 36$.

(a) $n(E) = 4$; namely, (4, 1), (3, 2), (2, 3), and (1, 4).

$$P(\text{sum of 5}) = \frac{n(E)}{n(S)} = \frac{4}{36} = \frac{1}{9}$$

(b) $n(E) = 8$; namely, (6, 1), (5, 2), (4, 3), (3, 4), (2, 5), (1, 6), (6, 5), and (5, 6).

$$P(\text{sum of 7 or 11}) = \frac{n(E)}{n(S)} = \frac{8}{36} = \frac{2}{9}$$

(c) The sum must be 10, 11, or 12. $n(E) = 6$; namely, (6, 4), (5, 5), (4, 6), (6, 5), (5, 6), and (6, 6).

$$P(\text{sum greater than 9}) = \frac{n(E)}{n(S)} = \frac{6}{36} = \frac{1}{6}$$

(d) The sum must be 2, 3, 4, or 5. $n(E) = 10$; namely, (1, 1), (2, 1), (1, 2), (3, 1), (2, 2), (1, 3), (4, 1), (3, 2), (2, 3), and (1, 4).

$$P(\text{sum less than or equal to 5}) = \frac{n(E)}{n(S)}$$
$$= \frac{10}{36}$$
$$= \frac{5}{18}$$

(e) $n(E) = 11$; namely, (3, 1), (3, 2), (3, 3), (3, 4), (3, 5), (3, 6), (1, 3), (2, 3), (4, 3), (5, 3), and (6, 3).

$$P(3 \text{ on one or both dice}) = \frac{n(E)}{n(S)} = \frac{11}{36}$$

(f) The sum must be 3, 5, 7, 9, or 11. $n(E) = 18$; namely, (2, 1), (1, 2), (4, 1), (3, 2), (2, 3), (1, 4), (6, 1), (5, 2), (4, 3), (3, 4), (2, 5), (1, 6), (6, 3), (5, 4), (4, 5), (3, 6), (6, 5), and (5, 6).

$$P(\text{sum that is odd}) = \frac{n(E)}{n(S)} = \frac{18}{36} = \frac{1}{2}$$

(g) There are only nine possibilities that do not have a prime number on one or both dice, namely, (1, 1), (1, 4), (1, 6), (4, 1), (4, 4), (4, 6), (6, 1), (6, 4), and (6, 6). So, $n(E) = 36 - 9 = 27$

$$P(\text{prime on one or both dice}) = \frac{n(E)}{n(S)}$$
$$= \frac{27}{36}$$
$$= \frac{3}{4}$$

(h) $n(E) = 36$, since all have a sum greater than 1.

$$P(\text{sum greater than 1}) = \frac{n(E)}{n(S)} = \frac{36}{36} = 1$$

17. Use branches (segments) emanating from one point to show possibilities for first stage (i.e., for first experiment); then use segments from the tip of each of these segments to show possibilities for second stage; and so on.

19. There are six possibilities for each die rolled. Since there are three dice rolled, there are $6 \cdot 6 \cdot 6 = 216$ different outcomes.

21. $P(\text{sum of 6}) = \dfrac{n(E)}{n(S)} = \dfrac{10}{216} = \dfrac{5}{108}$

Exercise Set 11-3

1. (a) There are three ways to get a sum of 10,

$$P(E) = \frac{3}{36} = \frac{1}{12}.$$

odds in favor $= \dfrac{P(E)}{1 - P(E)}$

$$= \frac{\frac{1}{12}}{1 - \frac{1}{12}}$$

$$= \frac{1}{12} \div \frac{11}{12}$$

$$= \frac{1}{12} \cdot \frac{12}{11}$$

$$= \frac{1}{11} \text{ or } 1{:}11$$

(b) There is one way to get a sum of 12,

$$P(E) = \frac{1}{36}.$$

odds in favor $= \dfrac{P(E)}{1 - P(E)}$

$$= \frac{\frac{1}{36}}{1 - \frac{1}{36}}$$

$$= \frac{1}{36} \div \frac{35}{36}$$

$$= \frac{1}{36} \cdot \frac{36}{35}$$

$$= \frac{1}{35} \text{ or } 1{:}35$$

(c) There are 30 ways to not get a sum of 7,

$$P(\overline{E}) = \frac{30}{36} = \frac{5}{6}.$$

odds against $= \dfrac{P(\overline{E})}{1 - P(\overline{E})}$

$$= \frac{\frac{5}{6}}{1 - \frac{5}{6}}$$

$$= \frac{5}{6} \div \frac{1}{6}$$

$$= \frac{5}{6} \cdot \frac{6}{1}$$

$$= \frac{5}{1} \text{ or } 5{:}1$$

(d) There are 34 ways to not get a sum of 3,

$$P(\overline{E}) = \frac{34}{36} = \frac{17}{18}.$$

odds against $= \dfrac{P(\overline{E})}{1 - P(\overline{E})}$

$$= \frac{\frac{17}{18}}{1 - \frac{17}{18}}$$

$$= \frac{17}{18} \div \frac{1}{18}$$

$$= \frac{17}{18} \cdot \frac{18}{1}$$

$$= \frac{17}{1} \text{ or } 17{:}1$$

(e) There are six ways to get doubles,

$$P(E) = \frac{6}{36} = \frac{1}{6}.$$

odds in favor $= \dfrac{P(E)}{1 - P(E)}$

$$= \frac{\frac{1}{6}}{1 - \frac{1}{6}}$$

$$= \frac{1}{6} \div \frac{5}{6}$$

$$= \frac{1}{6} \cdot \frac{6}{5}$$

$$= \frac{1}{5} \text{ or } 1{:}5$$

3. (a) There are four ways to get a queen,

$$P(E) = \frac{4}{52} = \frac{1}{13}.$$

odds in favor $= \dfrac{P(E)}{1 - P(E)}$

$$= \frac{\frac{1}{13}}{1 - \frac{1}{13}}$$

$$= \frac{1}{13} \div \frac{12}{13}$$

$$= \frac{1}{13} \cdot \frac{13}{12}$$

$$= \frac{1}{12} \text{ or } 1{:}12$$

(b) There are 12 ways to get a face card,

$$P(E) = \frac{12}{52} = \frac{3}{13}.$$

odds in favor $= \dfrac{P(E)}{1 - P(E)}$

$$= \frac{\frac{3}{13}}{1 - \frac{3}{13}}$$

$$= \frac{3}{13} \div \frac{10}{13}$$

$$= \frac{3}{13} \cdot \frac{13}{10}$$

$$= \frac{3}{10} \text{ or } 3{:}10$$

(c) There are 39 ways to not get a club,

$$P(\overline{E}) = \frac{39}{52} = \frac{3}{4}.$$

odds against $= \dfrac{P(\overline{E})}{1 - P(\overline{E})}$

$$= \frac{\frac{3}{4}}{1 - \frac{3}{4}}$$

$$= \frac{3}{4} \div \frac{1}{4}$$

$$= \frac{3}{4} \cdot \frac{4}{1}$$

$$= \frac{3}{1} \text{ or } 3{:}1$$

(d) There are four ways to get an ace,

$$P(E) = \frac{4}{52} = \frac{1}{13}.$$

odds in favor $= \dfrac{P(E)}{1 - P(E)}$

$$= \frac{\frac{1}{13}}{1 - \frac{1}{13}}$$

$$= \frac{1}{13} \div \frac{12}{13}$$

$$= \frac{1}{13} \cdot \frac{13}{12}$$

$$= \frac{1}{12} \text{ or } 1{:}12$$

(e) There are 26 ways to get a black card,

$$P(E) = \frac{26}{52} = \frac{1}{2}.$$

odds in favor $= \dfrac{P(E)}{1 - P(E)}$

$$= \frac{\frac{1}{2}}{1 - \frac{1}{2}}$$

$$= \frac{1}{2} \div \frac{1}{2}$$

$$= \frac{1}{2} \cdot \frac{2}{1}$$

$$= \frac{1}{1} \text{ or } 1{:}1$$

5. (a) $a = 7, b = 4$

$$P(E) = \frac{a}{a+b} = \frac{7}{7+4} = \frac{7}{11}$$

(b) $c = 2, d = 5$

$$P(\overline{E}) = \frac{c}{c+d} = \frac{2}{2+5} = \frac{2}{7}$$

$$P(E) = 1 - P(\overline{E}) = 1 - \frac{2}{7} = \frac{5}{7}$$

(c) $a = 3, b = 1$

$$P(E) = \frac{a}{a+b} = \frac{3}{3+1} = \frac{3}{4}$$

(d) $a = 1, d = 4$

$$P(\overline{E}) = \frac{c}{c+d} = \frac{1}{1+4} = \frac{1}{5}$$

$$P(E) = 1 - P(\overline{E}) = 1 - \frac{1}{5} = \frac{4}{5}$$

7. $c = 9, d = 5$

$$P(\overline{E}) = \frac{c}{c+d} = \frac{9}{9+5} = \frac{9}{14}$$

$$P(E) = 1 - P(\overline{E}) = 1 - \frac{9}{14} = \frac{5}{14}$$

9. $E(x) = \$4995 \cdot \dfrac{1}{2500} + (-\$5) \cdot \dfrac{2499}{2500} = -\$3.00$

or

$$E(x) = \$5000 \cdot \frac{1}{2500} - \$5 = -\$3.00$$

11. The probability of winning, getting doubles, is $\dfrac{1}{6}$. The expected value should be zero if the game is fair. Let X be the cost to play.

$$0 = \$5 \cdot \frac{1}{6} - X$$

$$X = \$\left(\frac{5}{6}\right) \text{ or about } \$0.83$$

13. $E(x) = \$1000 \cdot \dfrac{1}{1000} + \$500 \cdot \dfrac{1}{1000} + \$100 \cdot \dfrac{5}{1000} - \$3 = -\$1.00$

15. The probability of winning is $\dfrac{1}{1000}$.

$E(x) = \$500 \cdot \dfrac{1}{1000} - \$1 = -\$0.50$

The probability of winning when number 123 is played and boxed is $\dfrac{6}{1000}$ or $\dfrac{3}{500}$.

$E(x) = \$80 \cdot \dfrac{3}{500} - \$1 = -\$0.52$

17. Odds in favor of event E give an indication of the frequency for E to occur—called a "win." Example: If such odds are 4:11, then there would be, on average, a win four times out of 15 tries. Odds against entail a similar concept, but have to do with E not occurring—called a "loss." Example: if the odds against E are 8:5, then there would be, on average, eight losses out of 13 tries.

19. Odds are used to make a game mathematically fair, or to give one party a mathematical advantage over the other party. If, on average, a player has as much chance of winning an amount as of losing the *same* amount, then the game is considered fair.

21. It's a mathematically expected result or average, such as over the long run. More specifically: If the "parts" or outcomes of event E are X_1, X_2, ..., X_n, and the corresponding probabilities are $P(X_1)$ etc., then the expected value $X_1 P(X_1) + X_2 P(X_2) + \cdots + X_n P(X_n)$.

23. Recognize that once a 6 is rolled, the game is over. For the game involving four rolls, the probability of getting a 6 on the first roll is $\dfrac{1}{6}$. If he does not get a 6 on the first roll, the probability of getting a 1 through 5 on the first roll and a 6 on the second roll is $\dfrac{5}{6} \cdot \dfrac{1}{6}$. Continuing this process, the probability of getting a 6 in four rolls is

$\dfrac{1}{6} + \dfrac{5}{6} \cdot \dfrac{1}{6} + \left(\dfrac{5}{6}\right)^2 \cdot \dfrac{1}{6} + \left(\dfrac{5}{6}\right)^3 \cdot \dfrac{1}{6} = 0.518$. Using this same reasoning, the probability of getting two 6s in 24 rolls is 0.491. The probability for the gambler winning is larger on the first game.

Exercise Set 11-4

1. The events are mutually exclusive. There are a total of 12 months.
$P(\text{April or May}) = P(\text{April}) + P(\text{May})$

$\qquad = \dfrac{1}{12} + \dfrac{1}{12}$

$\qquad = \dfrac{2}{12}$

$\qquad = \dfrac{1}{6}$

3. The events are mutually exclusive. There are a total of $7 + 5 + 3 + 4 = 19$ instructors.
$P(\text{science or math}) = P(\text{science}) + P(\text{math})$

$\qquad = \dfrac{4}{19} + \dfrac{7}{19}$

$\qquad = \dfrac{11}{19}$

5. The events are mutually exclusive. There are a total of $8 + 9 + 3 = 20$ pens.

(a) $P(\text{red or blue}) = P(\text{red}) + P(\text{blue})$
$$= \frac{8}{20} + \frac{9}{20}$$
$$= \frac{17}{20}$$

(b) $P(\text{red or black}) = P(\text{red}) + P(\text{black})$
$$= \frac{8}{20} + \frac{3}{20}$$
$$= \frac{11}{20}$$

(c) $P(\text{blue or black}) = P(\text{blue}) + P(\text{black})$
$$= \frac{9}{20} + \frac{3}{20}$$
$$= \frac{12}{20} = \frac{3}{5}$$

7. The events are mutually exclusive. There are a total of $5 + 4 + 2 + 3 + 3 = 17$ professors.

(a) $P(\text{English or psychology})$
$= P(\text{English}) + P(\text{psychology})$
$$= \frac{5}{17} + \frac{3}{17}$$
$$= \frac{8}{17}$$

(b) $P(\text{mathematics or science})$
$= P(\text{mathematics}) + P(\text{science})$
$$= \frac{4}{17} + \frac{2}{17}$$
$$= \frac{6}{17}$$

(c) $P(\text{history, science, or mathematics})$
$= P(\text{history}) + P(\text{science}) + P(\text{mathematics})$
$$= \frac{3}{17} + \frac{2}{17} + \frac{4}{17}$$
$$= \frac{9}{17}$$

(d) $P(\text{English, mathematics, history})$
$= P(\text{English}) + P(\text{mathematics}) + P(\text{history})$
$$= \frac{5}{17} + \frac{4}{17} + \frac{3}{17}$$
$$= \frac{12}{17}$$

9.

Class	Females	Males	Total
Juniors	6	12	18
Seniors	6	4	10
Total	12	16	28

(a) $P(\text{junior or female}) = P(\text{junior}) + P(\text{female}) - P(\text{junior and female})$

$$= \frac{18}{28} + \frac{12}{28} - \frac{6}{28}$$

$$= \frac{24}{28} = \frac{6}{7}$$

(b) $P(\text{senior or female}) = P(\text{senior}) + P(\text{female}) - P(\text{senior and female})$

$$= \frac{10}{28} + \frac{12}{28} - \frac{6}{28}$$

$$= \frac{16}{28} = \frac{4}{7}$$

(c) $P(\text{junior or senior}) = P(\text{junior}) + P(\text{senior})$

$$= \frac{18}{28} + \frac{10}{28}$$

$$= \frac{28}{28} = 1$$

11.

Product	Company A	Company B	Company C	Total
Dresses	24	18	12	54
Blouses	13	36	15	64
Total	37	54	27	118

(a) $P(\text{A or dress}) = P(\text{A}) + P(\text{dress}) - P(\text{A and dress})$

$$= \frac{37}{118} + \frac{54}{118} - \frac{24}{118}$$

$$= \frac{67}{118}$$

(b) $P(\text{B or C}) = P(\text{B}) + P(\text{C})$

$$= \frac{54}{118} + \frac{27}{118}$$

$$= \frac{81}{118}$$

(c) $P(\text{blouse or A}) = P(\text{blouse}) + P(\text{A}) - P(\text{blouse and A})$

$$= \frac{64}{118} + \frac{37}{118} - \frac{13}{118}$$

$$= \frac{88}{118} = \frac{44}{59}$$

13.

Marital status	Cashiers	Stock clerks	Deli personnel	Total
Married	8	12	3	23
Not married	5	15	2	22
Total	13	27	5	45

(a) $P(\text{stock clerk or married}) = P(\text{stock clerk}) + P(\text{married}) - P(\text{stock clerk and married})$

$$= \frac{27}{45} + \frac{23}{45} - \frac{12}{45}$$

$$= \frac{38}{45}$$

(b) $P(\text{not married}) = 1 - P(\text{married})$

$$= 1 - \frac{23}{45}$$

$$= \frac{22}{45}$$

(c) $P(\text{cashier or not married}) = P(\text{cashier}) + P(\text{not married}) - P(\text{cashier and not married})$

$$= \frac{13}{45} + \frac{22}{45} - \frac{5}{45}$$

$$= \frac{30}{45} = \frac{2}{3}$$

15.

Type of show	Channel 6	Channel 8	Channel 10	Total
Quiz show	5	2	1	8
Comedy	3	2	8	13
Drama	4	4	2	10
Total	12	8	11	31

(a) $P(\text{quiz or 8}) = P(\text{quiz}) + P(8) - P(\text{quiz and 8})$

$$= \frac{8}{31} + \frac{8}{31} - \frac{2}{31}$$

$$= \frac{14}{31}$$

(b) $P(\text{drama or comedy}) = P(\text{drama}) + P(\text{comedy})$

$$= \frac{10}{31} + \frac{13}{31}$$

$$= \frac{23}{31}$$

(c) $P(\text{10 or drama}) = P(10) + P(\text{drama}) - P(\text{10 and drama})$

$$= \frac{11}{31} + \frac{10}{31} - \frac{2}{31}$$

$$= \frac{19}{31}$$

17. (a) $P(\text{king or queen or jack}) = P(\text{king}) + P(\text{queen}) + P(\text{jack})$

$$= \frac{4}{52} + \frac{4}{52} + \frac{4}{52}$$

$$= \frac{12}{52} = \frac{3}{13}$$

(b) $P(\text{club or heart or spade}) = P(\text{club}) + P(\text{heart}) + P(\text{spade})$

$$= \frac{13}{52} + \frac{13}{52} + \frac{13}{52}$$

$$= \frac{39}{52} = \frac{3}{4}$$

(c) $P(\text{king or queen or diamond})$

$= P(\text{king}) + P(\text{queen}) + P(\text{diamond}) - P(\text{king of diamonds or queen of diamonds})$

$$= \frac{4}{52} + \frac{4}{52} + \frac{13}{52} - \frac{2}{52}$$

$$= \frac{19}{52}$$

(d) $P(\text{ace or diamond or heart}) = P(\text{ace}) + P(\text{diamond}) + P(\text{heart}) - P(\text{ace of diamonds or ace of hearts})$

$$= \frac{4}{52} + \frac{13}{52} + \frac{13}{52} - \frac{2}{52}$$

$$= \frac{28}{52} = \frac{7}{13}$$

(e) $P(\text{9 or 10 or spade or club})$

$= P(9) + P(10) + P(\text{spade}) + P(\text{club}) - P(\text{9 of spades or clubs}) - P(\text{10 of spades or clubs})$

$$= \frac{4}{52} + \frac{4}{52} + \frac{13}{52} + \frac{13}{52} - \frac{2}{52} - \frac{2}{52}$$

$$= \frac{30}{52} = \frac{15}{26}$$

19. The events are mutually exclusive. There are a total of $6 + 2 + 1 + 1 = 10$ balls.

$P(\text{red or white}) = P(\text{red}) + P(\text{white})$

$$= \frac{6}{10} + \frac{1}{10}$$

$$= \frac{7}{10}$$

21. They are events that cannot occur at the same time.

23. $P(\text{mushrooms or pepperoni}) = P(\text{mushrooms}) + P(\text{pepperoni}) - P(\text{mushrooms and pepperoni})$

$0.43 = 0.32 + 0.17 - P(\text{mushrooms and pepperoni})$

$P(\text{mushrooms and pepperoni}) = 0.32 + 0.17 - 0.43 = 0.06$

25. $P(\text{not two-car}) = 1 - P(\text{two-car}) = 1 - 0.70 = 0.30$

Exercise Set 11-5

1. $P(\text{three underweight}) = (0.18)(0.18)(0.18)$
≈ 0.0058

3. $P(\text{four used seat belt}) = (0.52)(0.52)(0.52)(0.52)$
≈ 0.0073

5. $P(\text{two not citizens}) = (0.25)(0.25) = 0.0625$

7. $P(\text{two in same month}) = \frac{1}{12}$

Whatever the birth month of the first person, the probability that the second person was born in that same month is $\frac{1}{12}$.

9. $P(\text{three agree}) = \left(\frac{1}{2}\right)\left(\frac{1}{2}\right)\left(\frac{1}{2}\right) = \frac{1}{8}$

11. $P(D_1 \text{ and } D_2) = \frac{2}{6} \cdot \frac{1}{5} = \frac{2}{30} = \frac{1}{15}$

13. $P(\text{three without}) = (0.06)(0.06)(0.06) = 0.000216$

15. (a) $P(\text{three jacks}) = \dfrac{4}{52} \cdot \dfrac{3}{51} \cdot \dfrac{2}{50}$

$= \dfrac{24}{132,600}$

$= 0.00018$

(b) $P(\text{three clubs}) = \dfrac{13}{52} \cdot \dfrac{12}{51} \cdot \dfrac{11}{50}$

$= \dfrac{1716}{132,600}$

≈ 0.013

(c) $P(\text{three red}) = \dfrac{26}{52} \cdot \dfrac{25}{51} \cdot \dfrac{24}{50}$

$= \dfrac{15,600}{132,600}$

≈ 0.12

17. $P(\text{none pregnant}) = \dfrac{3}{8} \cdot \dfrac{2}{7} \cdot \dfrac{1}{6} = \dfrac{6}{336} = \dfrac{1}{56} \approx 0.018$

19. $P(3 \mid 4) = \dfrac{P(\text{H and 3})}{P(\text{H})} = \dfrac{\frac{1}{12}}{\frac{1}{2}} = \dfrac{1}{12} \cdot \dfrac{2}{1} = \dfrac{1}{6}$

21. $P(\text{sum 7} \mid \text{one 6}) = \dfrac{P(\text{one 6 and sum 7})}{P(\text{one 6})}$

$= \dfrac{\frac{2}{36}}{\frac{11}{36}}$

$= \dfrac{2}{36} \cdot \dfrac{36}{11}$

$= \dfrac{2}{11}$

23. $P(\text{diamond} \mid \text{face card})$

$= \dfrac{P(\text{face card and diamond})}{P(\text{face card})}$

$= \dfrac{\frac{3}{52}}{\frac{12}{52}}$

$= \dfrac{3}{52} \cdot \dfrac{52}{12}$

$= \dfrac{1}{4}$

25. $P(\text{red} \mid \text{off}) = \dfrac{P(\text{odd and red})}{P(\text{odd})} = \dfrac{\frac{2}{6}}{\frac{3}{6}} = \dfrac{2}{6} \cdot \dfrac{6}{3} = \dfrac{2}{3}$

27. $P(\text{less than } 5 \,|\, \text{red}) = \dfrac{P(\text{red and less than } 5)}{P(\text{red})}$

$$= \dfrac{\frac{3}{6}}{\frac{3}{6}}$$

$$= \dfrac{3}{6} \cdot \dfrac{6}{3}$$

$$= 1$$

29.

	Less than 200	200–399	400–599	600 or more	Total
Men	56	18	10	16	100
Women	61	18	13	8	100
Total	117	36	23	24	200

$P(\text{less than } 200 \,|\, \text{woman}) = \dfrac{P(\text{woman and less than } 200)}{P(\text{woman})}$

$$= \dfrac{\frac{61}{200}}{\frac{100}{200}}$$

$$= \dfrac{61}{200} \cdot \dfrac{200}{100}$$

$$= 0.61$$

31. $P(\text{woman} \,|\, 200 - 399)$

$$= \dfrac{P(200 - 399 \text{ and woman})}{P(200 - 399)}$$

$$= \dfrac{\frac{18}{200}}{\frac{36}{200}}$$

$$= \dfrac{18}{200} \cdot \dfrac{200}{36}$$

$$= \dfrac{1}{2}$$

33. Two events A and B are independent if the fact that A occurs has no effect on the probability of B occurring. They are dependent if such an effect does not exist. Examples of two recent events are
Independent: cutting your tree in your backyard, and choosing Brand B for your calculator
Dependent: Being on time for the bus, and being on time for the meeting.

35. There are $6 \cdot 6 \cdot 6 = 216$ ways to roll three dice and 25 ways to get a sum of 9.

$$P(\text{sum of } 9) = \dfrac{25}{216}$$

37. $P(\text{three same birthday}) = \left(\dfrac{1}{365}\right)^2$

$$= \dfrac{1}{133,225}$$

$$\approx (7.51)(10^{-6})$$

After the first person's birthday is revealed, the probability that the next person has the same is $\dfrac{1}{365}$ and the

probability that the last person has the same is $\dfrac{1}{365}$.

Exercise Set 11-6

1. $10! = 10 \cdot 9 \cdot 8 \cdot 7 \cdot 6 \cdot 5 \cdot 4 \cdot 3 \cdot 2 \cdot 1 = 3,628,800$

3. $9! = 9 \cdot 8 \cdot 7 \cdot 6 \cdot 5 \cdot 4 \cdot 3 \cdot 2 \cdot 1 = 362,880$

5. $0! = 1$

7. $\begin{aligned} _7P_5 &= \frac{7!}{(7-5)!} \\ &= \frac{7!}{2!} \\ &= \frac{7 \cdot 6 \cdot 5 \cdot 4 \cdot 3 \cdot 2!}{2!} \\ &= 7 \cdot 6 \cdot 5 \cdot 4 \cdot 3 \\ &= 2520 \end{aligned}$

9. $_5P_3 = \frac{5!}{(5-3)!} = \frac{5!}{2!} = \frac{5 \cdot 4 \cdot 3 \cdot 2!}{2!} = 5 \cdot 4 \cdot 3 = 60$

11. $_6P_0 = \frac{6!}{(6-0)!} = \frac{6!}{6!} = 1$

13. $_8P_8 = \frac{8!}{(8-8)!} = \frac{8!}{0!} = \frac{8 \cdot 7 \cdot 6 \cdot 5 \cdot 4 \cdot 3 \cdot 2 \cdot 1}{1} = 40,320$

15. $_6P_2 = \frac{6!}{(6-2)!} = \frac{6!}{4!} = \frac{6 \cdot 5 \cdot 4!}{4!} = 6 \cdot 5 = 30$

17. $_8P_3 = \frac{8!}{(8-3)!} = \frac{8!}{5!} = \frac{8 \cdot 7 \cdot 6 \cdot 5!}{5!} = 8 \cdot 7 \cdot 6 = 336$

19. $\begin{aligned} _7P_7 &= \frac{7!}{(7-7)!} \\ &= \frac{7!}{0!} \\ &= \frac{7!}{1} \\ &= 7 \cdot 6 \cdot 5 \cdot 4 \cdot 3 \cdot 2 \cdot 1 \\ &= 5040 \end{aligned}$

21. $\begin{aligned} _{20}P_5 &= \frac{20!}{(20-5)!} \\ &= \frac{20!}{15!} \\ &= \frac{20 \cdot 19 \cdot 18 \cdot 17 \cdot 16 \cdot 15!}{15!} \\ &= 20 \cdot 19 \cdot 18 \cdot 17 \cdot 16 \\ &= 1,860,480 \end{aligned}$

23. $\begin{aligned} _7P_5 &= \frac{7!}{(7-5)!} \\ &= \frac{7!}{2!} \\ &= \frac{7 \cdot 6 \cdot 5 \cdot 4 \cdot 3 \cdot 2!}{2!} \\ &= 7 \cdot 6 \cdot 5 \cdot 4 \cdot 3 \\ &= 2520 \end{aligned}$

25. $_5P_5 = \frac{5!}{(5-5)!} = \frac{5!}{0!} = \frac{5!}{1} = 5 \cdot 4 \cdot 3 \cdot 2 \cdot 1 = 120$

27. $\begin{aligned} _{50}P_6 &= \frac{50!}{(50-6)!} \\ &= \frac{50!}{44!} \\ &= \frac{50 \cdot 49 \cdot 48 \cdot 47 \cdot 46 \cdot 45 \cdot 44}{44!} \\ &= 11,441,304,000 \end{aligned}$

29. $4! = 4 \cdot 3 \cdot 2 \cdot 1 = 24$

31. Suppose, in a sequence of n events, are available:

k_1 possibilities for the first event;

k_2 possibilities for second event;

k_3 possibilities for the third; and so forth.

Then the total number of ways in which the sequence of n events can occur is

$k_1 \cdot k_2 \cdot k_3 \cdot \cdots \cdot k_n$.

33. (a) $\frac{n!}{k_1!k_2!\cdots k_p!} = \frac{11!}{1!4!4!2!} = 34,650$

(b) $\frac{n!}{k_1!k_2!\cdots k_p!} = \frac{10!}{3!2!1!4!} = 12,600$

Exercise Set 11-7

1. $_5C_2 = \frac{5!}{(5-2)!2!} = \frac{5!}{3!2!} = \frac{5 \cdot 4 \cdot 3!}{3! \cdot 2 \cdot 1} = 10$

3. $_7C_4 = \frac{7!}{(7-4)!4!} = \frac{7!}{3!4!} = \frac{7 \cdot 6 \cdot 5 \cdot 4!}{3 \cdot 2 \cdot 1 \cdot 4!} = 35$

5. $_6C_4 = \frac{6!}{(6-4)!4!} = \frac{6!}{2!4!} = \frac{6 \cdot 5 \cdot 4!}{2 \cdot 1 \cdot 4!} = 15$

7. $_3C_3 = \frac{3!}{(3-3)!3!} = \frac{3!}{0!3!} = 1$

9. $_{12}C_2 = \dfrac{12!}{(12-2)!2!} = \dfrac{12!}{10!2!} = \dfrac{12 \cdot 11 \cdot 10!}{10! \cdot 2 \cdot 1} = 66$

11. $_{52}C_5 = \dfrac{52!}{(52-5)!5!}$

$= \dfrac{52!}{47!5!}$

$= \dfrac{52 \cdot 51 \cdot 50 \cdot 49 \cdot 48 \cdot 47!}{47! \cdot 5 \cdot 4 \cdot 3 \cdot 2 \cdot 1}$

$= 2,598,960$

13. $_9C_5 = \dfrac{9!}{(9-5)!5!} = \dfrac{9!}{4!5!} = \dfrac{9 \cdot 8 \cdot 7 \cdot 6 \cdot 5!}{4 \cdot 3 \cdot 2 \cdot 1 \cdot 5!} = 126$

How many ways are there to choose three questions from the remaining seven?

$_7C_5 = \dfrac{7!}{(7-3)!3!} = \dfrac{7!}{(4!3!)} = \dfrac{7 \cdot 6 \cdot 5 \cdot 4!}{(4!3 \cdot 2 \cdot 1)} = 35$

15. $_{10}C_3 = \dfrac{10!}{(10-3)!3!} = \dfrac{10!}{7!3!} = \dfrac{10 \cdot 9 \cdot 8 \cdot 7!}{7! \cdot 3 \cdot 2 \cdot 1} = 120$

17. $_{11}C_6 = \dfrac{11!}{(11-6)!6!}$

$= \dfrac{11!}{5!6!}$

$= \dfrac{11 \cdot 10 \cdot 9 \cdot 8 \cdot 7 \cdot 6!}{5 \cdot 4 \cdot 3 \cdot 2 \cdot 1 \cdot 6!}$

$= 462$

19. $_4C_2 \cdot {}_{12}C_5 \cdot {}_7C_3 = \dfrac{4!}{2!2!} \cdot \dfrac{12!}{7!5!} \cdot \dfrac{7!}{4!3!}$

$= 6 \cdot 792 \cdot 35$

$= 166,320$

21. $_{10}C_3 \cdot {}_{10}C_3 = \dfrac{10!}{7!3!} \cdot \dfrac{10!}{7!3!} = 120 \cdot 120 = 14,400$

23. $_8C_2 \cdot {}_6C_3 \cdot {}_{10}C_3 = \dfrac{8!}{6!2!} \cdot \dfrac{6!}{3!3!} \cdot \dfrac{10!}{7!3!}$

$= 28 \cdot 20 \cdot 120$

$= 67,200$

25. $_{25}C_5 = \dfrac{25!}{20!5!} = \dfrac{25 \cdot 24 \cdot 23 \cdot 22 \cdot 21 \cdot 20!}{20! \cdot 5 \cdot 4 \cdot 3 \cdot 2 \cdot 1} = 53,130$

27. $_9C_5 = \dfrac{9!}{4!5!} = \dfrac{9 \cdot 8 \cdot 7 \cdot 6 \cdot 5!}{4 \cdot 3 \cdot 2 \cdot 1 \cdot 5!} = 126$

29. $_{30}C_{20} = \dfrac{30!}{10!20!} = 30,045,015$

31. A combination is a selection of objects without regard to order or arrangement.

33.

n\r	0	1	2	3	4	5	6
0	1						
1	1	1					
2	1	2	1				
3	1	3	3	1			
4	1	4	6	4	1		
5	1	5	10	10	5	1	
6	1	6	15	20	15	6	1

$_5C_0 = \dfrac{5!}{5!0!} = 1$ $_6C_0 = \dfrac{6!}{6!0!} = 1$

$_5C_1 = \dfrac{5!}{4!1!} = 5$ $_6C_1 = \dfrac{6!}{5!1!} = 6$

$_5C_2 = \dfrac{5!}{3!2!} = 10$ $_6C_2 = \dfrac{6!}{4!2!} = 15$

$_5C_3 = \dfrac{5!}{2!3!} = 10$ $_6C_3 = \dfrac{6!}{3!3!} = 20$

$_5C_4 = \dfrac{5!}{1!4!} = 5$ $_6C_4 = \dfrac{6!}{2!4!} = 15$

$_5C_5 = \dfrac{5!}{0!5!} = 1$ $_6C_5 = \dfrac{6!}{1!5!} = 6$

$_6C_6 = \dfrac{6!}{0!6!} = 1$

Exercise Set 11-8

1. Find the total number of ways to choose four people from 25.

$_{25}C_4 = \dfrac{25!}{21!4!} = 13,650$

(a) Find the number of ways to select all teachers.

$_5C_4 = \dfrac{5!}{1!4!} = 5$

The probability is $\dfrac{5}{12,650} = \dfrac{1}{2530} \approx 0.0004$.

(b) Find the number of ways to choose two teachers and two parents.

$_5C_2 \cdot {}_{20}C_2 = \dfrac{5!}{3!2!} \cdot \dfrac{20!}{18!2!} = 1900$

The probability is $\dfrac{1900}{12,650} = \dfrac{38}{253} \approx 0.15$.

(c) Find the number of ways to choose all parents.

$$_{20}C_4 = \frac{20!}{16!4!} = 4845$$

The probability is $\dfrac{4845}{12,650} = \dfrac{969}{2530} \approx 0.383$.

(d) Find the number of ways to choose one teacher and three parents.

$$_5C_1 \cdot {}_{20}C_3 = \frac{5!}{4!1!} \cdot \frac{20!}{17!3!} = 5700$$

The probability is $\dfrac{5700}{12,650} = \dfrac{114}{253} \approx 0.45$.

3. Find the total number of ways to select three people from ten.

$$_{10}C_3 = \frac{10!}{7!3!} = 120$$

(a) Find the number of ways to select all Republicans.

$$_4C_3 = \frac{4!}{1!3!} = 4$$

The probability is $\dfrac{4}{120} = \dfrac{1}{30} \approx 0.033$.

(b) Find the number of ways to select all Democrats.

$$_3C_3 = \frac{3!}{0!3!} = 1$$

The probability is $\dfrac{1}{120} \approx 0.0083$.

(c) Find the number of ways to select one of each.

$$_4C_1 \cdot {}_5C_1 \cdot {}_3C_1 = \frac{4!}{3!1!} \cdot \frac{3!}{2!1!} \cdot \frac{3!}{2!1!} = 36$$

The probability is $\dfrac{36}{120} = \dfrac{3}{10}$.

(d) Find the number of ways to select two Democrats and one Independent.

$$_3C_2 \cdot {}_3C_1 = \frac{3!}{1!2!} \cdot \frac{3!}{2!1!} = 9$$

The probability is $\dfrac{9}{120} = \dfrac{3}{40}$.

(e) Find the number of ways to select one Independent and two Republicans.

$$_3C_1 \cdot {}_4C_2 = \frac{3!}{2!1!} \cdot \frac{4!}{2!2!} = 18$$

The probability is $\dfrac{18}{120} = \dfrac{3}{20}$.

5. Find the total number of ways to select four from 12.

$$_{12}C_4 = \frac{12!}{8!4!} = 495$$

(a) Find the number of ways to select no defectives, or select four good resistors from nine.

$$_9C_4 = \frac{9!}{5!4!} = 126$$

The probability is $\dfrac{126}{495} = \dfrac{14}{55} \approx 0.25$.

(b) Find the number of ways to select one defective from three defectives and three good resistors from nine.

$$_3C_1 \cdot {}_9C_3 = \frac{3!}{2!1!} \cdot \frac{9!}{6!3!} = 252$$

The probability is $\dfrac{252}{495} = \dfrac{28}{55} \approx 0.51$.

(c) Find the number of ways to select three defective from three and one good resistor from nine.

$$_3C_3 \cdot {}_9C_1 = \frac{3!}{0!3!} \cdot \frac{9!}{8!1!} = 9$$

The probability is $\dfrac{9}{495} = \dfrac{1}{55} \approx 0.018$.

7. Find the total number of ways to choose five cards from 52.

$$_{52}C_5 = \frac{52!}{5!47!} = 2,598,960$$

Find the number of ways to select kinds in order.

$$_{13}P_2 = \frac{13!}{11!} = 156$$

Find the number of ways to select 3 out of 4 of the first kind.

$$_4C_3 = \frac{4!}{1!3!} = 4$$

Find the number of ways to select 2 out of 4 of the second kind.

$$_4C_2 = \frac{4!}{2!2!} = 6$$

The probability is $\dfrac{156 \cdot 4 \cdot 6}{2,598,960} = 0.00144$.

9. Find the total number of ways to select three from 15.

$$_{15}C_3 = \frac{15!}{12!3!} = 455$$

(a) Find the number of ways to select all life.

$$_8C_3 = \frac{8!}{5!3!} = 56$$

The probability is $\frac{56}{455} = \frac{8}{65} \approx 0.123$.

(b) Find the number of ways to select two homeowners and one of the others.

$$_2C_2 \cdot {_8C_1} + {_2C_2} \cdot {_5C_1}$$
$$= \frac{2!}{0!2!} \cdot \frac{8!}{7!1!} + \frac{2!}{0!2!} \cdot \frac{5!}{4!1!}$$
$$= 8 + 5$$
$$= 13$$

The probability is $\frac{13}{455} = \frac{1}{35} \approx 0.029$.

(c) Find the number of ways to select all automobile.

$$_5C_3 = \frac{5!}{2!3!} = 10$$

The probability is $\frac{10}{455} = \frac{2}{91} \approx 0.022$.

(d) Find the number of ways to choose one of each.

$$_8C_1 \cdot {_5C_1} \cdot {_2C_1} = \frac{8!}{7!1!} \cdot \frac{5!}{4!1!} \cdot \frac{2!}{1!1!} = 8 \cdot 5 \cdot 2 = 80$$

The probability is $\frac{80}{455} = \frac{16}{91} \approx 0.176$.

(e) Find the number of ways to choose two life and one automobile.

$$_8C_2 \cdot {_5C_1} = \frac{8!}{6!2!} \cdot \frac{5!}{4!1!} = 28 \cdot 5 = 140$$

The probability is $\frac{140}{455} = \frac{4}{13} \approx 0.308$.

11. Find the number of ways to select three science books from eight and four math books from nine.

$$_8C_3 \cdot {_9C_4} = \frac{8!}{5!3!} \cdot \frac{9!}{5!4!} = 7056$$

Find the total number of ways to select seven books from 17.

$$_{17}C_7 = \frac{17!}{10!7!} = 19,448$$

The probability is $\frac{7056}{19,448} = \frac{882}{2431} \approx 0.363$.

13. The probability is

$$\frac{1}{_{40}C_5} = \frac{1}{\frac{40!}{35!5!}} = \frac{1}{658,008} \approx 1.5 \times 10^{-6}.$$

15. Method A: From Section 11-5, use multiplication rule 2, extended to more than two events:

$$P(\text{woman, man, woman, man, woman})$$
$$= \frac{3}{5} \cdot \frac{2}{4} \cdot \frac{2}{3} \cdot \frac{1}{2} \cdot \frac{1}{1}$$
$$= \frac{12}{20}$$
$$= \frac{1}{10}.$$

Method B: From Exercise Set 11-6, use the formula given in Exercise 33 for the number of permutations of n objects, where k_1 are alike, k_2 are alike, etc. In the present case, $k_1 = 3$, $k_2 = 2$, and $k_1 + k_2 = n = 5$. Then the number of permutations is

$$\frac{n!}{k_1!k_2!} = \frac{5!}{3!2!} = \frac{120}{12} = 10.$$ Since only one of

these 10 meets our order, the desired

probability $= \frac{1}{10}$.

17. Find the number of ways to select four aces from four and the other card from 48.

$$_4C_4 \cdot {_{48}C_1} = 1 \cdot 48 = 48$$

Find the number of ways to select five cards from 52.

$$_{52}C_5 = \frac{52!}{47!5!} = 2,598,960$$

The probability is $\frac{48}{2,598,960} = \frac{1}{54,145}$.

19. There is only one way to obtain a royal flush in each suit and there are 4 suits. The probability is

$$\frac{1 \cdot 4}{_{52}C_5} = \frac{4}{2,598,960} = \frac{1}{649,740}.$$

Review Exercises

1. The sample space is {1, 2, 3, 4, 5, 6}. Since there are six outcomes, $n(S) = 6$.

(a) $P(5) = \frac{n(E)}{n(S)} = \frac{1}{6}$

(b) $P(6) = \frac{n(E)}{n(S)} = \frac{1}{6}$

(c) $P(\text{number less than 5}) = \frac{n(E)}{n(S)} = \frac{4}{6} = \frac{2}{3}$

3.

Juice	Frequency
orange	20
grapefruit	16
apple	9
Total	45

$$P(\text{grapefruit}) = \frac{f}{n} = \frac{16}{45}$$

5. $P(\text{cordless}) = \dfrac{f}{n} = \dfrac{850}{1500} = \dfrac{17}{30}$

7.

Color	Frequency
white	16
red	4
blue	3
yellow	7
Total	30

(a) $P(\text{blue}) = \dfrac{f}{n} = \dfrac{3}{30} = 0.1$

(b) $P(\text{yellow or red}) = \dfrac{f}{n} = \dfrac{11}{30}$

(c) $P(\text{white, yellow, or blue}) = \dfrac{F}{n} = \dfrac{26}{30} = \dfrac{13}{15}$

(d) $P(\text{not red}) = 1 - P(\text{red}) = 1 - \dfrac{4}{30} = \dfrac{26}{30} = \dfrac{13}{15}$

or

$P(\text{not red}) = P(\text{white, yellow, or blue}) = \dfrac{13}{15}$

(from part c.)

9. $P(\text{sum 8}\,|\,\text{one 5}) = \dfrac{P(\text{one 5 and sum 8})}{P(\text{one 5})}$

$$= \frac{\frac{2}{36}}{\frac{11}{36}}$$

$$= \frac{2}{36} \cdot \frac{36}{11}$$

$$= \frac{2}{11}$$

11. $P(\text{five worry}) = (0.78)(0.78)(0.78)(0.78)(0.78)$
≈ 0.289

13. (a) $P(\text{three black}) = \dfrac{26}{52} \cdot \dfrac{25}{51} \cdot \dfrac{24}{50}$

$$= \frac{15,600}{132,600}$$

$$= \frac{2}{17} \approx 0.118$$

(b) $P(\text{three spades}) = \dfrac{13}{52} \cdot \dfrac{12}{51} \cdot \dfrac{11}{50}$

$$= \frac{1716}{132,600}$$

$$= \frac{11}{850} \approx 0.013$$

(c) $P(\text{three queens}) = \dfrac{4}{52} \cdot \dfrac{3}{51} \cdot \dfrac{2}{50}$

$$= \frac{24}{132,600}$$

$$= \frac{1}{5525} \approx 1.81 \times 10^{-4}$$

15. The events are mutually exclusive. There are a total of $6 + 3 + 2 + 2 = 13$ pieces.
$P(\text{caramel or peppermint})$
$= P(\text{caramel}) + P(\text{peppermint})$
$= \dfrac{2}{13} + \dfrac{3}{13}$
$= \dfrac{5}{13}$

17. $P(\overline{E}) = 1 - P(E) = 1 - \dfrac{5}{6} = \dfrac{1}{6}$

$\text{odds against} = \dfrac{P(\overline{E})}{1 - P(\overline{E})}$

$$= \frac{\frac{1}{6}}{1 - \frac{1}{6}}$$

$$= \frac{\frac{1}{6}}{\frac{5}{6}}$$

$$= \frac{1}{6} \cdot \frac{6}{5}$$

$$= \frac{1}{5} \text{ or } 1{:}5$$

19. $E(x) = 1¢ \cdot \dfrac{1}{5} + 5¢ \cdot \dfrac{1}{5} + 10¢ \cdot \dfrac{1}{5} + 25¢ \cdot \dfrac{1}{5} + 50¢ \cdot \dfrac{1}{5}$
$= 18.2¢$

21. If repetitions are allowed,
number $= 26 \cdot 26 \cdot 26 \cdot 10 \cdot 10 \cdot 10 \cdot 10$
$= 175,760,000.$

If repetitions are allowed in letters only,
number $= 26 \cdot 26 \cdot 26 \cdot 10 \cdot 9 \cdot 8 \cdot 7 = 88,583,040.$

23. There are $5 \cdot 4 \cdot 3 = 60$ combinations.
No, a combination lock uses permutations, i.e.,

$$_5 P_3 = \frac{5!}{(5-3)!} = \frac{5!}{2!} = 5 \cdot 4 \cdot 3 = 60.$$

25. There are eight possibilities for the roll and two possibilities for the flip. The sample space is listed in the right-most column.

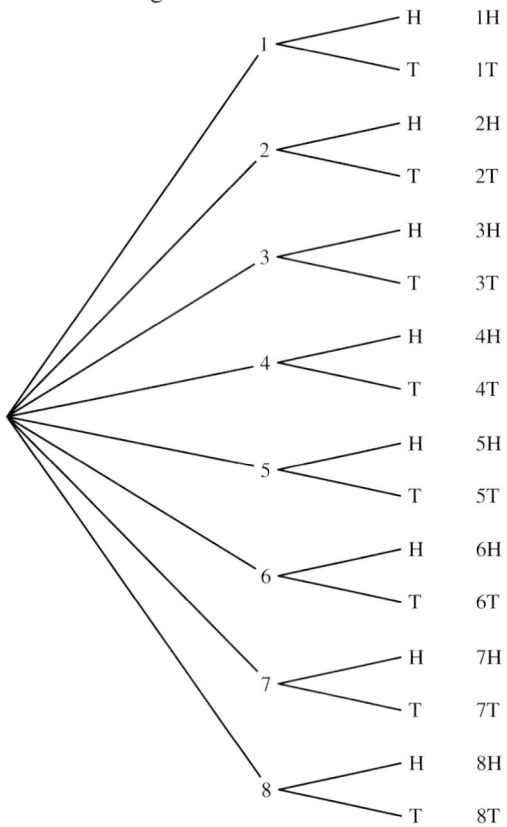

27. $_{12} C_5 = \dfrac{12!}{(12-5)!5!}$

$= \dfrac{12!}{7!5!}$

$= \dfrac{12 \cdot 11 \cdot 10 \cdot 9 \cdot 8 \cdot 7!}{7! \cdot 5 \cdot 4 \cdot 3 \cdot 2 \cdot 1}$

$= 792$

29. $_6 C_4 = \dfrac{6!}{(6-4)!4!} = \dfrac{6!}{2!4!} = \dfrac{6 \cdot 5 \cdot 4!}{2 \cdot 1 \cdot 4!} = 15$

31. Find the number of ways to select one of each.

$$_6 C_1 \cdot _3 C_1 \cdot _2 C_1 = \frac{6!}{5!1!} \cdot \frac{3!}{2!1!} \cdot \frac{2!}{1!1!} = 6 \cdot 3 \cdot 2 = 36$$

Find the number of ways to choose three investments from a total of 11.

$$_{11} C_3 = \frac{11!}{8!3!} = 165$$

The probability is $\dfrac{36}{165} = \dfrac{12}{55}.$

Chapter Test

1. Since there are 52 outcomes in the sample space, $n(S) = 52$.

   (a) $P(\text{jack}) = \dfrac{n(E)}{n(S)} = \dfrac{4}{52} = \dfrac{1}{13}$

   (b) $P(4) = \dfrac{n(E)}{n(S)} = \dfrac{4}{52} = \dfrac{1}{13}$

   (c) $P(\text{less than 6}) = \dfrac{n(E)}{n(S)} = \dfrac{16}{52} = \dfrac{4}{13}$

3.

Color	Frequency
blue	12
green	8
gray	4
black	7
Total	31

   (a) $P(\text{blue}) = \dfrac{f}{n} = \dfrac{12}{31}$

   (b) $P(\text{green or gray}) = \dfrac{f}{n} = \dfrac{12}{31}$

   (c) $P(\text{green or black or blue}) = \dfrac{f}{n} = \dfrac{27}{31}$

   (d) $P(\text{not black}) = 1 - P(\text{black}) = 1 - \dfrac{7}{31} = \dfrac{24}{31}$ or

        $P(\text{not black}) = P(\text{green or blue or gray})$

$$= \frac{f}{n}$$

$$= \frac{24}{31}$$

5. The sample space is listed in the right-most column.

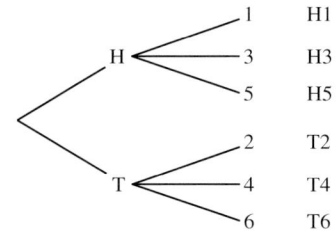

7. (a) $P(5 \text{ reds}) = \dfrac{26}{52} \cdot \dfrac{25}{51} \cdot \dfrac{24}{50} \cdot \dfrac{23}{49} \cdot \dfrac{22}{48}$

$= \dfrac{7,893,600}{311,875,200}$

≈ 0.025

(b) $P(5 \text{ diamonds}) = \dfrac{13}{52} \cdot \dfrac{12}{51} \cdot \dfrac{11}{50} \cdot \dfrac{10}{49} \cdot \dfrac{9}{48}$

$= \dfrac{154,440}{311,875,200}$

≈ 0.000495

(c) $P(5 \text{ aces}) = 0$ (there are only 4 aces per deck)

9. $P(\text{diamond} \mid \text{red}) = \dfrac{P(\text{red and diamond})}{P(\text{red})}$

$= \dfrac{\frac{13}{52}}{\frac{26}{52}}$

$= \dfrac{13}{52} \cdot \dfrac{52}{26}$

$= \dfrac{1}{2}$

11. $P(\text{H} \mid \text{even}) = \dfrac{P(\text{even and H})}{P(\text{even})} = \dfrac{\frac{1}{4}}{\frac{1}{2}} = \dfrac{1}{4} \cdot \dfrac{2}{1} = \dfrac{1}{2}$

13. There are 2 choices for each of the 25 questions.
$2^{25} = 33,554,432$

15. There are four possibilities for the first city and two possibilities for the second city. There are a total of 8 possibilities.

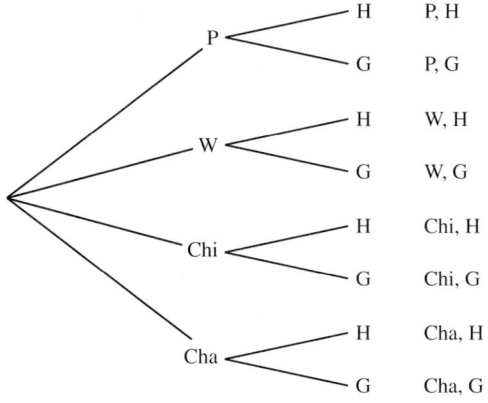

P	H	P, H
	G	P, G
W	H	W, H
	G	W, G
Chi	H	Chi, H
	G	Chi, G
Cha	H	Cha, H
	G	Cha, G

17. $_8P_8 = \dfrac{8!}{(8-8)!} = \dfrac{8!}{0!} = 8 \cdot 7 \cdot 6 \cdot 5 \cdot 4 \cdot 3 \cdot 2 \cdot 1 = 40,320$

19. $_{15}C_4 = \dfrac{15!}{11!4!} = \dfrac{15 \cdot 14 \cdot 13 \cdot 12 \cdot 11!}{11! \cdot 4 \cdot 3 \cdot 2 \cdot 1} = 1365$

21. There are $_2C_1$ possibilities for a top and $_2C_1$ possibilities for bottoms.
$_2C_1 \cdot _2C_1 = 2 \cdot 2 = 4$

23. $P(\overline{E}) = 1 - P(E) = 1 - \dfrac{4}{9} = \dfrac{5}{9}$

odds against $= \dfrac{P(\overline{E})}{1 - P(\overline{E})}$

$= \dfrac{\frac{5}{9}}{1 - \frac{5}{9}}$

$= \dfrac{\frac{5}{9}}{\frac{4}{9}}$

$= \dfrac{5}{9} \cdot \dfrac{9}{4}$

$= \dfrac{5}{4}$ or 5:4

25. $E(x) = \dfrac{1}{6} \cdot 4 + \dfrac{1}{6} \cdot 5 + \dfrac{1}{6} \cdot 2 + \dfrac{1}{6} \cdot 10 + \dfrac{1}{6} \cdot 3 + \dfrac{1}{6} \cdot 7$

$= 5\dfrac{1}{6}$

27. Find the number of ways to select one of each.
$_5C_1 \cdot _4C_1 \cdot _2C_1 = 5 \cdot 4 \cdot 2 = 40$
Find the number of ways to select three from a total of 11.

$_{11}C_3 = \dfrac{11!}{8!3!} = 165$

The probability is $\dfrac{40}{165} = \dfrac{8}{33}$.

12 Statistics

Exercise Set 12-1

1.

Rank	Tally	Frequency
Fr	卌 卌 卌 ///	18
So	卌 卌 //	12
Jr	卌 /	6
Se	////	4

3. highest value − lowest value = 416 − 21 = 395

$$\frac{\text{difference}}{\text{number of classes}} = \frac{395}{6} = 65.8 \approx 66$$

Start with lowest value and add 66 to get the lower class limits: 21, 87, 153, 219, 285, 351. Set up the classes by subtracting one from each lower class limit except the first lower class limit.

Class	Tally	Frequency
21–86	//	2
87–152	//	2
153–218	卌 卌 //	12
219–284	卌 卌 卌 ///	18
285–350	卌 ////	9
351–416	卌 /	6

5. highest value − lowest value = 4040 − 70 = 3970

$$\frac{\text{difference}}{\text{number of classes}} = \frac{3970}{7} = 567.1 \approx 568$$

Start with the lowest value and add 568 to get the lower class limits: 70, 638, 1206, 1774, 2342, 2910, 3478. Set up the classes by subtracting one from each lower class limit except the first lower class limit.

Class	Tally	Frequency
70–637	卌 卌 卌	15
638–1205	卌 /	6
1206–1773	///	3
1774–2341		0
2342–2909	/	1
2910–3477		0
3478–4045	//	2

7. highest value − lowest value = 775 − 5 = 770

$$\frac{\text{difference}}{\text{number of classes}} = \frac{770}{8} = 96.25 \approx 97$$

Start withy the lowest value and add 97 to get the lower limits: 5, 102, 199, 296, 393, 490, 587, 684. Set up the classes by subtracting one from each lower class limit except the first lower class limit.

Class	Tally	Frequency
5–101	卌 卌 卌 //	17
102–198	卌 /	6
199–295	卌 /	6
296–392	//	2
393–489	//	2
490–586	///	3
587–683	/	1
684–780	//	2

9. combined highest − combined lowest

= 550 − 306

= 244

$$\frac{\text{difference}}{\text{number of classes}} = \frac{244}{8} = 30.5 \approx 31$$

Start with the lowest value and add 31 to get the lower class limits: 306, 337, 368, 399, 430, 461, 492, 523.
Set up the classes by subtracting one from each lower class limit except the first lower class limit.

Class	McGwire: Tally & Frequency		Sosa: Tally & Frequency	
306–336	/	1		0
337–367	ⅢⅡ /	6	ⅢⅡ ⅢⅡ	10
368–398	ⅢⅡ ⅢⅡ ⅢⅡ ////	19	ⅢⅡ ⅢⅡ ⅢⅡ /	16
399–429	ⅢⅡ ⅢⅡ ⅢⅡ	15	ⅢⅡ ⅢⅡ ⅢⅡ ⅢⅡ /	21
430–460	ⅢⅡ ⅢⅡ ⅢⅡ ///	18	ⅢⅡ ⅢⅡ ⅢⅡ	15
461–491	ⅢⅡ /	6	///	3
492–522	///	3	/	1
523–553	//	2		0

11. Arrange the data in order. (Note 3, 8, and 9 are written as 03, 08, and 09.) Separate the data according to the first digit. Use the first digits as stems and the second digits as leaves.

Stems	Leaves
0	3 8 9 9
1	0 2 2 2 4 4 4 6 8 9
2	1 2 2 5 8 8
3	1 3 6 7
4	1 9
5	2 4 8

Analysis: Ten executives make 10–19 calls, while six made 21–28 calls.

13. Arrange the data in order. Separate the data according to the first digit. Use the first digits as stems and the second digits as leaves.

Stems	Leaves
1	2
2	0 3
3	2 5 8 8 9
4	1 3 3
5	0 1 2 3 3 5 8 9 9

15. Measurements or observations that are gathered for an event under study are called data.

17. A population consists of all subjects under study; a sample is a representative subgroup, or subset, of the population.

19. Number each member of the population, and then select every kth member. The starting number, though, must be selected at random.

21. Have an intact group of subjects that represent the population.

23. (a) It is a cluster sample since an intact group of subjects that represent the population is used for a sample.

(b) It is a systematic sample since every seventh customer is selected.

(c) It is a random sample since each subject of the population has an equal chance of being selected.

(d) It is a systematic sample since every hundredth hamburger is checked.

(e) It is a stratified sample since the population is divided into groups and members from each group are randomly selected.

25. Answers will vary.

Exercise Set 12-2

1. Draw the bars with heights corresponding to the number.

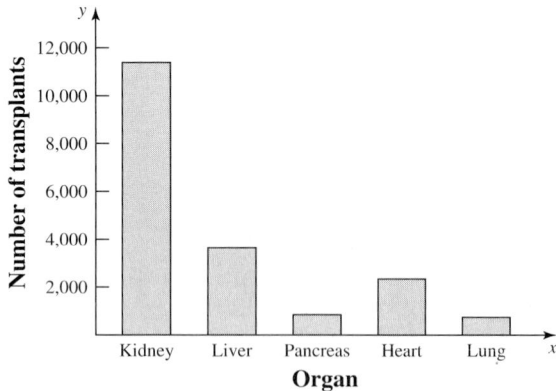

3. Draw the bars with heights corresponding to the number.

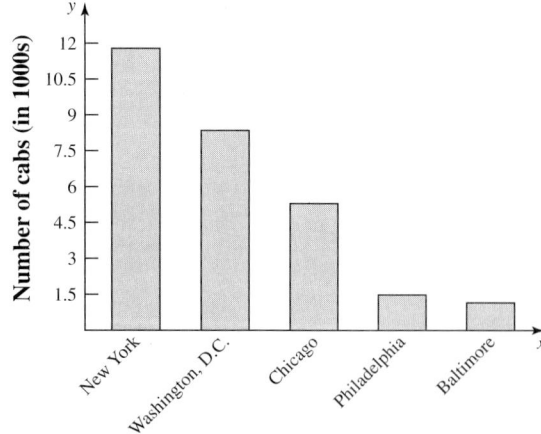

5.

Rank	Frequency, f	degrees $= \frac{f}{n} \cdot 360°$	percent $= \frac{f}{n} \cdot 100\%$
Freshmen	12	$\frac{12}{90} \cdot 360° = 48°$	$\frac{12}{90} \cdot 100\% \approx 13\%$
Sophomores	25	$\frac{25}{90} \cdot 360° = 100°$	$\frac{25}{90} \cdot 100\% \approx 28\%$
Juniors	36	$\frac{36}{90} \cdot 360° = 144°$	$\frac{36}{90} \cdot 100\% = 40\%$
Seniors	17	$\frac{17}{90} \cdot 360° = 68°$	$\frac{17}{90} \cdot 100\% \approx 19\%$

$n = 90$

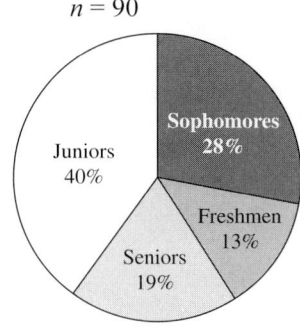

7.

Reason	Number f	degrees $= \frac{f}{n} \cdot 360°$	percent $= \frac{f}{n} \cdot 100\%$
To support self/family	62	$\frac{62}{100} \cdot 360° = 223.2°$	$\frac{62}{100} \cdot 100\% = 62\%$
For extra money	18	$\frac{18}{100} \cdot 360° = 64.8°$	$\frac{18}{100} \cdot 100\% = 18\%$
For something different	12	$\frac{12}{100} \cdot 360° = 43.2°$	$\frac{12}{100} \cdot 100\% = 12\%$
Other	8	$\frac{8}{100} \cdot 360° = 28.8°$	$\frac{8}{100} \cdot 100\% = 8\%$

$n = 100$

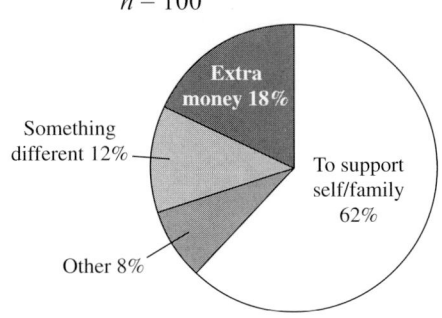

9. For the histogram, draw vertical bars corresponding to the frequencies for each class.

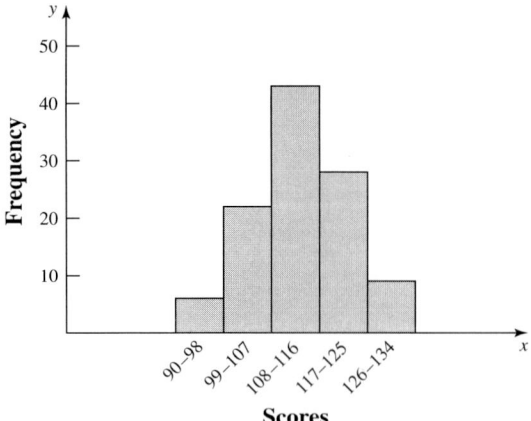

For the frequency polygon, find the midpoints for each class: 94, 103, 112, 121, and 130. Label the horizontal axis with the midpoints. Connect adjacent midpoints with straight lines. Finish the graph by drawing a line back to the horizontal at the beginning and end of graph.

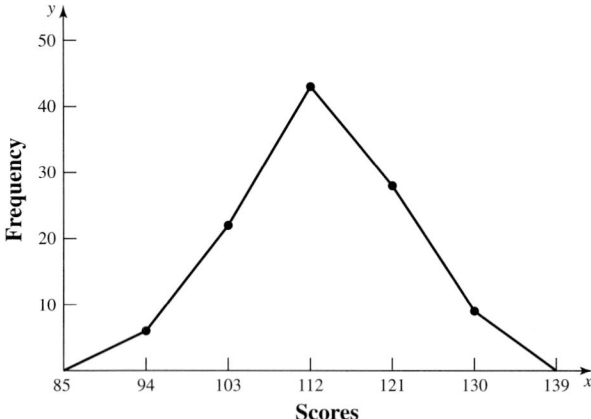

11. For the histogram, draw vertical bars corresponding to the frequencies for each class.

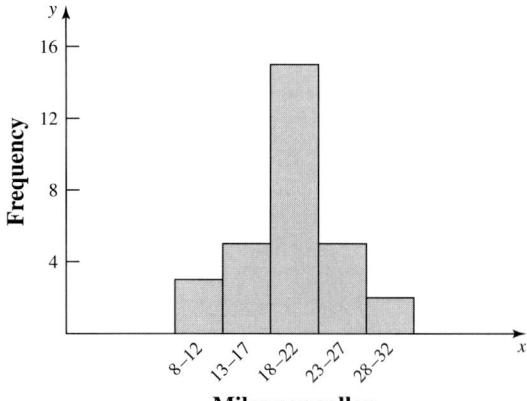

For the frequency polygon, find the midpoints for each class: 10, 15, 20, 25, and 30. Label the horizontal axis with the midpoints. Connect adjacent midpoints with straight lines. Finish the graph by drawing a line back to the horizontal at the beginning and end of graph.

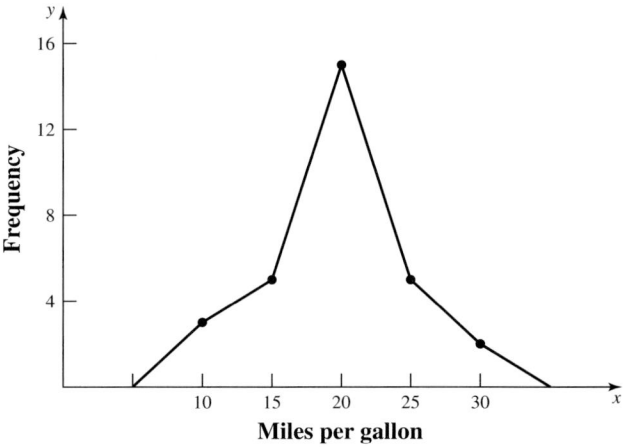

13. Represent the years on the *x* axis and the number on the *y* axis, and then draw straight lines through the points.

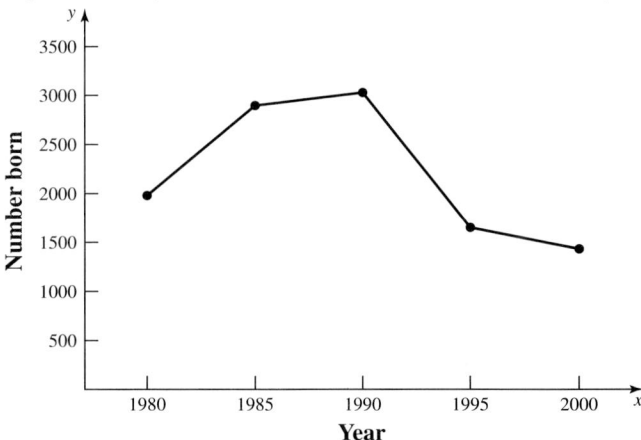

15. Represent the years on the *x* axis and the books on the *y* axis, and then draw straight lines through the points.

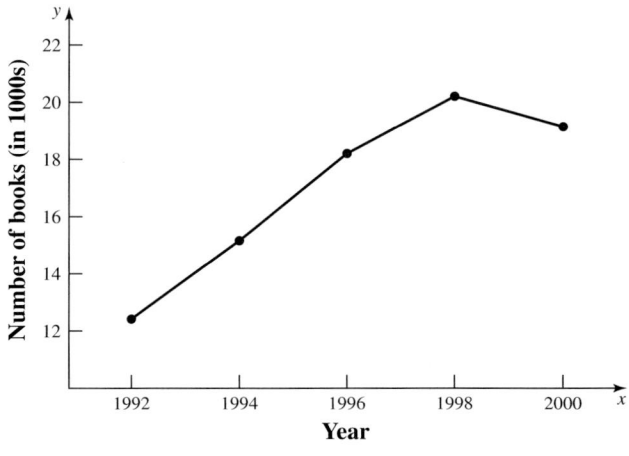

17. *Relationship*: Both indicate the frequency for each class.
Difference: The histogram represents each entire class by means of rectangles/bars touching their neighbors. In the frequency polygon, we use the midpoints of each class, and connect the "frequency points" with straight segments.

19. (a) Time series graph

   (b) Pie graph

   (c) Bar graph

(d) Time series graph

(e) Pie graph

(f) Bar graph

Exercise Set 12-3

1. $\text{mean} = \overline{X} = \dfrac{\sum X}{n} = \dfrac{61+11+1+3+2+30+18+3+7}{9}$

$$= \dfrac{136}{9}$$

$$\approx 15.11$$

Arrange the data in order.
1 2 3 3 7 11 18 30 61
The median is the middle value, MD = 7.
The value that occurs most often is 3, mode = 3.

$\text{MR} = \dfrac{L+H}{2} = \dfrac{1+61}{2} = 31$

3. $\text{mean} = \overline{X} = \dfrac{\sum X}{n} = \dfrac{700+298+638+260+1380+280+270+1350+380+570}{10}$

$$= \dfrac{6126}{10}$$

$$= 612.6$$

Since the data are in thousands, the mean is 612,600. Arrange the data in order.
260 270 280 298 380 570 638 700 1350 1380

The median is the middle point, $\text{MD} = \dfrac{380+570}{2} = 475$ or 475,000. Since each value occurs only once, there is

no mode.

$\text{MR} = \dfrac{L+H}{2} = \dfrac{260+1380}{2} = 820$ or 820,000

5. All answers are in billions.

$\text{mean} = \overline{X} = \dfrac{\sum X}{n} = \dfrac{516.5+540.3+\cdots+478.6}{8}$

$$= \dfrac{4316.4}{8}$$

$$= 539.55$$

Arrange the data in order.
416.7 478.6 516.5 540.3 560.7 584.7 603.6 615.3

The median is the middle point, $\text{MD} = \dfrac{540.3+560.7}{2} = 550.5.$

Since each value occurs only once, there is no mode.

$\text{MR} = \dfrac{L+H}{2} = \dfrac{416.7+615.3}{2} = 516$

7. $\text{mean} = \overline{X} = \dfrac{\sum X}{n} = \dfrac{340+75+123+259+151}{5}$

$$= \dfrac{948}{5}$$

$$= 189.6$$

Arrange the data in order.
75 123 151 259 340
The median is the middle value, MD = 151.

Exercise Set 12-4

1. $R = \text{highest value} - \text{lowest value} = 61 - 1 = 60$

$$\overline{X} = \frac{\sum X}{n} = \frac{61 + 1 + \cdots + 7}{9} = \frac{126}{9} = 14$$

X	$X - \overline{X}$	$(X - \overline{X})^2$
61	47	2209
1	−13	169
1	−13	169
3	−11	211
2	−12	144
30	16	256
18	4	16
3	−11	121
7	−7	49
		3254

$$s^2 = \frac{\sum (X - \overline{X})^2}{n - 1} = \frac{3254}{8} = 406.75$$

$$x = \sqrt{s^2} = \sqrt{406.75} \approx 20.17$$

Since there is one extremely high value (61), and since s takes into account all values, s is the best measure in this case.

3. $R = \text{highest value} - \text{lowest value}$
$= 1902 - 103$
$= 1799$

$$\overline{X} = \frac{\sum X}{n} = \frac{1902 + 103 + \cdots 656}{10} = \frac{7166}{10} = 716.6$$

X	$X - \overline{X}$	$(X - \overline{X})^2$
1902	1185.4	1,405,173.16
103	−613.6	376,504.96
653	−63.6	4044.96
1901	1184.4	1,402,803.36
788	71.4	5097.96
361	−355.6	126,451.36
216	−500.6	250.600.36
363	−353.6	125,032.96
223	−493.6	243,640.96
656	−60.6	3672.36
		3,943,022.4

$$s^2 = \frac{\sum (X - \overline{X})^2}{n - 1} = \frac{3,943,022.4}{9} = 438,113.6$$

$$s = \sqrt{s^2} = \sqrt{438,113.6} \approx 661.90$$

5. $R = \text{highest value} - \text{lowest value}$
$= 305 - 206$
$= 99$

$$\overline{X} = \frac{\sum X}{n} = \frac{206 + 215 + \cdots + 261}{9}$$
$$= \frac{2387}{9}$$
$$\approx 265.2$$

X	$X - \overline{X}$	$(X - \overline{X})^2$
206	−59.2	3504.64
215	−50.2	2520.04
305	39.8	1584.04
297	31.8	1011.24
265	−0.2	0.04
282	16.8	282.24
301	35.8	1281.64
255	−10.2	104.04
261	−4.2	17.64
		10,305.56

$$s^2 = \frac{\sum (X - \overline{X})^2}{n - 1} = \frac{10,305.56}{8} = 1288.195$$

$$s = \sqrt{s^2} = \sqrt{1288.195} \approx 35.89$$

7. $R = \text{highest value} - \text{lowest value} = 78 - 68 = 10$

$$\overline{X} = \frac{\sum X}{n} = \frac{78 + 72 + \cdots + 72}{9} = \frac{654}{9} \approx 72.7$$

X	$X - \overline{X}$	$(X - \overline{X})^2$
78	5.3	28.09
72	−0.7	0.49
68	−4.7	22.09
73	0.3	0.09
75	2.3	5.29
69	−3.7	13.69
74	1.3	1.69
73	0.3	0.09
72	−0.7	0.49
		72.01

$$s^2 = \frac{\sum(X - \overline{X})^2}{n-1} = \frac{72.01}{8} \approx 9$$

$$s = \sqrt{s^2} = \sqrt{9} = 3$$

9. R = highest value – lowest value

$\quad$ = \$3.80 – \$3.20

$\quad$ = \$0.60

$$\overline{X} = \frac{\sum X}{n} = \frac{3.80 + 3.80 + \cdots + 3.69}{7}$$

$$= \frac{25.42}{7}$$

$$\approx 3.63$$

X	$X - \overline{X}$	$(X - \overline{X})^2$
3.80	0.17	0.0289
3.80	0.17	0.0289
3.20	–0.43	0.1849
3.57	–0.06	0.0036
3.62	–0.01	0.0001
3.74	0.11	0.0121
3.69	0.06	0.0036
		0.2621

$$s^2 = \frac{\sum(X - \overline{X})^2}{n-1} = \frac{0.2621}{6} \approx 0.04$$

$$s = \sqrt{s^2} = \sqrt{0.044} \approx 0.2$$

11. R = highest value – lowest value

$\quad$ = 1380 – 260

$\quad$ = 1120 or 1,120,000

$$\overline{X} = \frac{\sum X}{n} = \frac{700 + 298 + \cdots + 570}{10}$$

$$= \frac{6126}{10}$$

$$= 612.6 \text{ or } 612,600$$

X	$X - \overline{X}$	$(X - \overline{X})^2$
700	87.4	7638.76
298	–314.6	98,973.16
638	25.4	645.16
260	–352.6	124,326.76
1380	767.4	588,902.76
280	–332.6	110,622.76
270	–342.6	117,374.76
1350	737.4	543,758.76
380	–232.6	54,102.76
570	–42.6	1814.76
		1,648,160.4

$$s^2 = \frac{\sum(X - \overline{X})^2}{n-1}$$

$$= \frac{1,648,160.4}{9}$$

$$\approx 183,128.9333 \text{ or } 183,128,933.3$$

$$s = \sqrt{s^2} = \sqrt{183,128,933.3}$$

$$\approx 427.93566 \text{ or } 427,935.66$$

13. Range, variance, and standard deviation

15. Because (a) the range uses only two of the values in the data set, and (b) an extremely large, and/or an extremely low, value can make the range very large—thus giving the impression of more variability than is actually the case.

17. Find the mean and subtract it from each value in the data set. Square each difference, and find the sum of all such differences. Divide this sum by $(n - 1)$, where n = number of values in data set. Take the square root of the quotient to obtain the standard deviation.

19. The variation is not the same. Find the standard deviation for each data set.

(a) $R = 12$; $\overline{X} = 11$

X	$X - \overline{X}$	$(X - \overline{X})^2$
5	−6	36
7	−4	16
9	−2	4
11	0	0
13	2	4
15	4	16
17	6	36
		112

$$s^2 = \frac{112}{6} \approx 18.7; \quad s = \sqrt{18.7} \approx 4.32$$

(b) $R = 12$; $\overline{X} = 11$

X	$X - \overline{X}$	$(X - \overline{X})^2$
5	−6	36
6	−5	25
7	−4	16
11	0	0
15	4	16
16	5	25
17	6	36
		154

$$s^2 = \frac{154}{6} \approx 25.7; \quad s = \sqrt{25.7} \approx 5.07$$

(c) $R = 12$; $\overline{X} = 11$

X	$X - \overline{X}$	$(X - \overline{X})^2$
5	−6	36
5	−6	36
5	−6	36
11	0	0
17	6	36
17	6	36
17	6	36
		216

$$s^2 = \frac{216}{6} = 36; \quad s = \sqrt{36} = 6$$

Exercise Set 12-5

1. Arrange the 20 data values in order.
24 27 29 30 32 33 35 36 37 38
38 40 41 42 43 44 47 48 49 50

(a) There are 4 values below 32.

$$\frac{4}{20} = 0.20 = 20\%$$

A score of 32 is equivalent to the 20th percentile.

(b) There are 15 values below 44.

$$\frac{15}{20} = 0.75 = 75\%$$

A score of 44 is equivalent to the 75th percentile.

(c) There are 7 values below 36.

$$\frac{7}{20} = 0.35 = 35\%$$

A score of 36 is equivalent to the 35th percentile.

(d) There is 1 value below 27.

$$\frac{1}{20} = 0.05 = 5\%$$

A score of 27 is equivalent to the 5th percentile.

(e) There are 18 values below 49.

$$\frac{18}{20} = 0.90 = 90\%$$

A score of 49 is equivalent to the 90th percentile.

3. There are $500 - 125 = 375$ values below Carveta's rank.

$$\frac{375}{500} = 0.75 = 75\% \text{ or 75th percentile}$$

5. There are $200 - 43 = 157$ places after her.

$$\frac{157}{200} = 0.785 = 78.5\% \text{ or } 79\%, \text{ or } (78.5 \text{ or } 79)\text{th}$$

percentile

7. 20% of $50 = 0.2 \times 50 = 10$
Hence, 10 students scored lower than she.

9. Find Bill's percentile rank and compare it to 60%.

$600 - 220 = 380$

$$\frac{380}{600} \approx 0.63 = 63\%$$

Bill scored higher because he is at the 63rd percentile.

11. Arrange the 20 data values in order.
18 19 20 21 23 24 24 27 27 28
32 32 33 34 35 37 37 42 43 43

(a) There are 12 values below 33.

$$\frac{12}{20} = 0.60 = 60\%$$

The age of 33 is equivalent to the 60th percentile.

(b) 8, since 33 is the 8th value from the top.

(c) 20% of $20 = 0.20 \times 20 = 4$
Hence, an age of 23 corresponds to the 20th percentile because there are 4 values below it.

13. Arrange the data in order.
20 21 24 29 34 48 52 55 59
The median is the middle value.
$$Q_2 = 34$$
Find the median of the values less than Q_2.
$$Q_1 = \frac{21 + 24}{2} = 22.5$$
Find the median of the values above Q_2.
$$Q_3 = \frac{52 + 55}{2} = 53.5$$

15. Arrange the data in order.
278 310 327 352 390 407 539 583 883 952
The median is the middle point.
$$Q_2 = \frac{390 + 407}{2} = 398.5$$
Find the median of the values less than Q_2.
$$Q_1 = 327$$
Find the median of the values above Q_2.
$$Q_3 = 583$$

17. A percentile is a percentage that equals the percent of data values lying below a certain given point.

19. No, they are not the same. The *numerical class rank* tells how far down the list, starting from the top, a person or object is (Thus, "first" gives rank 1, "second" gives rank 2, etc.). The *percentile rank* is the percentage of values lying below a certain point—that is, below a certain value—in the data set.

21. The 90th percentile would be higher.

23. About 50th percentile

Exercise Set 12-6

1.

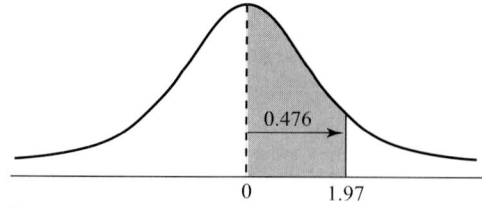

The area between $z = 0$ and $z = 1.97$ is 0.476.

3.

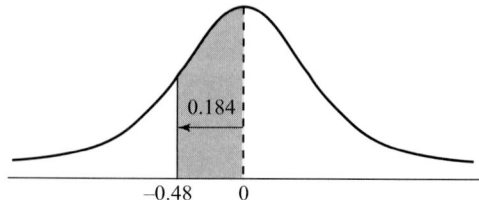

The area between $z = 0$ and $z = -0.48$ is 0.184.

5.

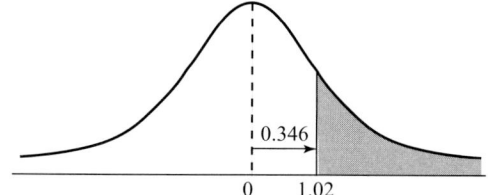

The area to the right of $z = 1.02$ is
$0.500 - 0.346 = 0.154$.

7.

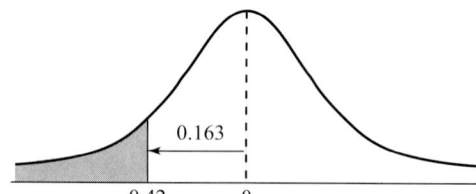

The area to the left of $z = -0.42$ is
$0.500 - 0.163 = 0.337$.

9.

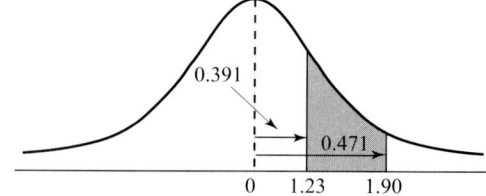

The area between $z - 1.23$ and $z = 1.90$ is
$0.471 - 0.391 = 0.08$.

11.

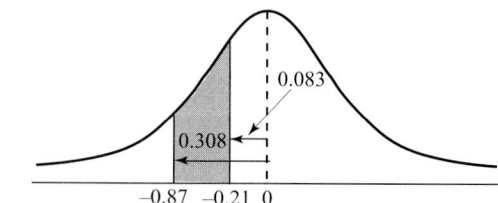

The area between $z = -0.87$ and $z = -0.21$ is
$0.308 - 0.083 = 0.225$.

13.

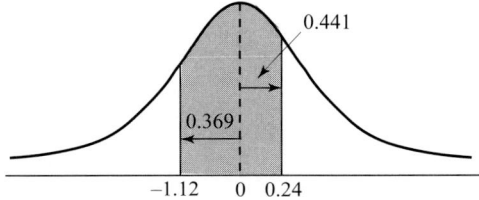

The area between $z = 0.24$ and $z = -1.12$ is
$0.095 + 0.369 = 0.464$.

15.

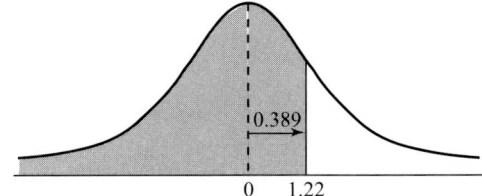

The area to the left of $z = 1.22$ is
$0.500 + 0.389 = 0.889$.

17.

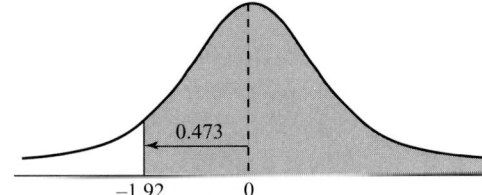

The area to the right of $z = -1.92$ is
$0.500 + 0.473 = 0.973$.

19. The normal distribution is bell-shaped, unimodal, symmetrical about the mean, and continuous. Furthermore,

(a) It never touches the x-axis.

(b) The area under the curve is 1, and is divided approximately as
(i) 0.68 within 1 standard deviation of the mean
(ii) 0.95 within 2 standard deviations of the mean
(iii) 0.997 within 3 standard deviations of the mean

21. 1

23. Find a z value for which the area between it and 0 is $0.5398 - 0.500 = 0.0398 \approx 0.040$. From Appendix C, $z = +0.100$.

25. (a) $1.00 - 0.05 = 0.95$

$\dfrac{0.95}{2} = 0.475$

From Appendix C, $z = \pm 1.96$.

(b) $1.00 - 0.10 = 0.90$

$\dfrac{0.90}{2} = 0.450$

From Appendix C, $z = \pm 1.64$.

(c) $1.00 - 0.01 = 0.99$

$\dfrac{0.99}{2} = 0.495$

From Appendix C, $z = 2.58$.

Exercise Set 12-7

1. (a) $z = \dfrac{\text{value} - \text{mean}}{\text{standard deviation}}$

$= \dfrac{12.55 - 11.76}{2.72}$

$= 0.29$

The area between $z = 0$ and $z = 0.29$ is 0.114. Since the desired area is in the right tail, subtract 0.114 from 0.500.

$0.500 - 0.114 = 0.386$

Hence, the probability that a randomly selected production worker earns more than $12.55 is 0.386, or 38.6%.

(b) $z = \dfrac{\text{value} - \text{mean}}{\text{standard deviation}}$

$= \dfrac{8.00 - 11.76}{2.72}$

$= -1.38$

The area between $z = 0$ and $z = -1.38$ is 0.416. Since the desired area is in the left tail, subtract 0.416 from 0.500.

$0.500 - 0.416 = 0.084$

Hence, the probability that a randomly selected production worker earns less than $8.00 is 0.084 or 8.4%.

3. (a) $z = \dfrac{\text{value} - \text{mean}}{\text{standard deviation}}$

$= \dfrac{35,000 - 29,835}{3000}$

$= 1.72$

The area between $z = 0$ and $z = 1.72$ is 0.457. Since the desired area is in the right tail, subtract 0.457 from 0.500.

$0.500 - 0.457 = 0.043$

Hence, the probability that a teacher earns more than $35,000 is 0.043, or 4.3%.

(b) $z = \dfrac{\text{value} - \text{mean}}{\text{standard deviation}}$

$= \dfrac{25,000 - 29,835}{3000}$

$= -1.61$

The area between $z = 0$ and $z = -1.61$ is 0.446. Since the desired area is in the left tail, subtract 0.446 from 0.500.

$0.500 - 0.446 = 0.054$

Hence, the probability that the teacher earns less than $25,000 is 0.054, or 5.4%.

5. (a) $z = \dfrac{\text{value} - \text{mean}}{\text{standard deviation}} = \dfrac{2.5 - 4.8}{0.89} = -2.58$

The area between $z = 0$ and $z = -2.58$ is 0.495. Since the desired area is in the left tail, subtract 0.495 from 0.500.
$0.500 - 0.495 = 0.005$
Hence, the probability that he or she owned the set less than 2.5 years is 0.005, or 0.5%.

(b) $z_1 = \dfrac{\text{value} - \text{mean}}{\text{standard deviation}} = \dfrac{3 - 4.8}{0.89} = -2.02$

$z_2 = \dfrac{\text{value} - \text{mean}}{\text{standard deviation}} = \dfrac{4 - 4.8}{0.89} = -0.90$

The area between $z = 0$ and $z = -2.02$ is 0.478. The area between $z = 0$ and $z = -0.90$ is 0.316. Since the desired area is between $z = -2.02$ and $z = -0.90$, subtract 0.316 from 0.478.
$0.478 - 0.316 = 0.162$
Hence, the probability that he or she owned the set between 3 and 4 years is 0.162, or 16.2%.

(c) $z = \dfrac{\text{value} - \text{mean}}{\text{standard deviation}} = \dfrac{4.2 - 4.8}{0.89} = -0.67$

The area between $z = 0$ and $z = -0.67$ is 0.249. Since the area under the normal curve to the right of $z = -0.67$ is desired, add 0.500 to 0.249.
$0.500 + 0.249 = 0.749$
Hence, the probability that he or she owned the set more than 4.2 years is 0.749, or 74.9%.

7. (a) $z_1 = \dfrac{\text{value} - \text{mean}}{\text{standard deviation}}$
$= \dfrac{25,000 - 30,000}{2000}$
$= -2.5$

$z_2 = \dfrac{\text{value} - \text{mean}}{\text{standard deviation}}$
$= \dfrac{28,000 - 30,000}{2000}$
$= -1$

The area between $z = 0$ and $z = -2.5$ is 0.494. The area between $z = 0$ and $z = -1$ is 0.341. Since the desired area is between $z = -1$ and $z = -2.5$, subtract 0.341 from 0.494.
$0.494 - 0.341 = 0.153$
Hence, the probability that a tire's lifetime is between 25,000 and 28,000 miles is 0.153, or 15.3%.

(b) $z_1 = \dfrac{\text{value} - \text{mean}}{\text{standard deviation}}$
$= \dfrac{27,000 - 30,000}{2000}$
$= -1.5$

$z_2 = \dfrac{\text{value} - \text{mean}}{\text{standard deviation}}$
$= \dfrac{32,000 - 30,000}{2000}$
$= 1$

The area between $z = 0$ and $z = -1.5$ is 0.433. The area between $z = 0$ and $z = 1$ is 0.341. The total area is
$0.433 + 0.341 = 0.774$.
Hence, the probability that a tire's lifetime is between 27,000 and 32,000 miles is 0.774, or 77.4%.

(c) $z_1 = \dfrac{\text{value} - \text{mean}}{\text{standard deviation}}$
$= \dfrac{31,500 - 30,000}{2000}$
$= 0.75$

$z_2 = \dfrac{\text{value} - \text{mean}}{\text{standard deviation}}$
$= \dfrac{33,500 - 30,000}{2000}$
$= 1.75$

The area between $z = 0$ and $z = 0.75$ is 0.273. The area between $z = 0$ and $z = 1.75$ is 0.460. Since the desired area is between $z = 0.75$ and $z = 1.75$, subtract 0.273 from 0.460.
$0.460 - 0.273 = 0.187$
Hence, the probability that a tire's lifetime is between 31,500 and 33,500 miles is 0.187, or 18.7%.

9. (a) $z = \dfrac{\text{value} - \text{mean}}{\text{standard deviation}} = \dfrac{50 - 44}{6} = 1$

The area between $z = 0$ and $z = 1$ is 0.341. Since the desired area is to the left of $z = 1$, add 0.500 to 0.341.
$0.500 + 0.341 = 0.841$
Hence, the probability that at most 50 inches of snow will be received is 0.841, or 84.1%.

(b) $z = \dfrac{\text{value} - \text{mean}}{\text{standard deviation}} = \dfrac{53 - 44}{6} = 1.5$

The area between $z = 0$ and $z = 1.5$ is 0.433. Since the desired area is in the left tail, subtract 0.433 from 0.500.
$0.500 - 0.433 = 0.067$
Hence, the probability that at least 53 inches of snow will be received is 0.067, or 6.7%.

11. (a) $z_1 = \dfrac{\text{value} - \text{mean}}{\text{standard deviation}} = \dfrac{15 - 24.6}{5.8} = -1.66$

$z_2 = \dfrac{\text{value} - \text{mean}}{\text{standard deviation}} = \dfrac{30 - 24.6}{5.8} = 0.93$

The area between $z = 0$ and $z = -1.66$ is 0.452. The area between $z = 0$ and $z = 0.93$ is 0.324. The total area is $0.452 + 0.324 = 0.776$. Hence, the probability that it will take a student between 15 and 30 minutes to complete the test is 0.776, or 77.6%.

(b) $z_1 = \dfrac{\text{value} - \text{mean}}{\text{standard deviation}} = \dfrac{18 - 24.6}{5.8} = -1.14$

$z_2 = \dfrac{\text{value} - \text{mean}}{\text{standard deviation}} = \dfrac{28 - 24.6}{5.8} = 0.59$

The area between $z = 0$ and $z = -1.14$ is 0.373. Since the desired area is to the left of $z = -1.14$, subtract 0.373 from 0.500.
$0.500 - 0.373 = 0.127$
The area between $z = 0$ and $z = 0.59$ is 0.222. Since the desired area is to the right of $z = 0.59$, subtract 0.222 from 0.500.
$0.500 - 0.222 = 0.278$
The total desired area is
$0.127 + 0.278 = 0.405$. Hence, the probability that it will take a student less than 18 minutes or more than 28 minutes to complete the test is 0.405, or 40.5%.

13. (a) $z = \dfrac{\text{value} - \text{mean}}{\text{standard deviation}} = \dfrac{62 - 64.2}{3.2} = -0.69$

The area between $z = 0$ and $z = -0.69$ is 0.255. Since the desired area is to the right of $z = -0.69$, add 0.500 to 0.255.
$0.500 + 0.255 = 0.755$
Hence, the probability that the temperature will be above 62° is 0.755, or 75.5%.

(b) $z = \dfrac{\text{value} - \text{mean}}{\text{standard deviation}} = \dfrac{67 - 64.2}{3.2} = 0.88$

The area between $z = 0$ and $z = 0.88$ is 0.311. Since the desired area is to the left of $z = 0.88$, add 0.500 to 0.311.
$0.500 + 0.311 = 0.811$
Hence, the probability that the temperature will be below 67° is 0.811, or 81.1%.

(c) $z_1 = \dfrac{\text{value} - \text{mean}}{\text{standard deviation}} = \dfrac{65 - 64.2}{3.2} = 0.25$

$z_2 = \dfrac{\text{value} - \text{mean}}{\text{standard deviation}} = \dfrac{68 - 64.2}{3.2} = 1.19$

The area between $z = 0$ and $z = 0.25$ is 0.099. The area between $z = 0$ and $z = 1.19$ is 0.383. Since the desired area is between $z = 0.25$ and $z = 1.19$, subtract 0.099 from

0.383.
$0.383 - 0.099 = 0.284$
Hence, the probability that the temperature will be between 65° and 68° is 0.284, or 28.4%.

15. (a) $z = \dfrac{\text{value} - \text{mean}}{\text{standard deviation}} = \dfrac{93 - 100}{15} = -0.47$

The area between $z = 0$ and $z = -0.47$ is 0.181. Since the desired area is in the left tail, subtract 0.181 from 0.500.
$0.500 - 0.181 = 0.319$
$0.319 \times 2000 = 638$
Hence, 638 people will score below 93.

(b) $z = \dfrac{\text{value} - \text{mean}}{\text{standard deviation}} = \dfrac{120 - 100}{15} = 1.33$

The area between $z = 0$ and $z = 1.33$ is 0.408. Since the desired area is in the right tail, subtract 0.408 from 0.500.
$0.500 - 0.408 = 0.092$
$0.092 \times 2000 = 184$
Hence, 184 people will score above 120.

(c) $z_1 = \dfrac{\text{value} - \text{mean}}{\text{standard deviation}} = \dfrac{80 - 100}{15} = -1.33$

$z_2 = \dfrac{\text{value} - \text{mean}}{\text{standard deviation}} = \dfrac{105 - 100}{15} = 0.33$

The area between $z = 0$ and $z = -1.33$ is 0.408. The area between $z = 0$ and $z = 0.33$ is 0.129. The total area is
$0.408 + 0.129 = 0.537$.
$0.537 \times 2000 = 1074$
Hence 1074 people will score between 80 and 105.

(d) $z_1 = \dfrac{\text{value} - \text{mean}}{\text{standard deviation}} = \dfrac{75 - 100}{15} = -1.67$

$z_2 = \dfrac{\text{value} - \text{mean}}{\text{standard deviation}} = \dfrac{82 - 100}{15} = -1.20$

The area between $z = 0$ and $z = -1.67$ is 0.453. The area between $z = 0$ and $z = -1.20$ is 0.385. Since the desired area is between -1.67 and -1.20, subtract 0.385 from 0.453.
$0.453 - 0.385 = 0.068$
$0.068 \times 2000 = 136$
Hence, 136 people will score between 75 and 82.

17. (a) $z = \dfrac{\text{value} - \text{mean}}{\text{standard deviation}}$

$= \dfrac{150,000 - 145,500}{1500}$

$= 3$

The area between $z = 0$ and $z = 3$ is 0.499. Since the desired area is in the right tail, subtract 0.499 from 0.500.

$0.500 - 0.499 = 0.001$

$0.001 \times 800 = 0.8 \approx 1$

Hence, 1 home will cost more than \$150,000.

(b) $z_1 = \dfrac{\text{value} - \text{mean}}{\text{standard deviation}}$

$= \dfrac{141,000 - 145,500}{1500}$

$= -3$

$z_2 = \dfrac{\text{value} - \text{mean}}{\text{standard deviation}}$

$= \dfrac{151,000 - 145,500}{1500}$

$= 3.67$

The area between $z = 0$ and $z = -3$ is 0.499. The area between $z = 0$ and $z = 3.67$ is about 0.500. The desired area is $0.499 + 0.500 = 0.999$.

$0.999 \times 800 = 799.2 \approx 799$

Hence, 799 homes will cost between \$141,000 and \$151,000.

(c) $z = \dfrac{\text{value} - \text{mean}}{\text{standard deviation}}$

$= \dfrac{147,500 - 145,500}{1500}$

$= 1.33$

The area between $z = 0$ and $z = 1.33$ is 0.408. Since the desired area is to the left of $z = 1.33$, add 0.500 to 0.408.

$0.500 + 0.408 = 0.908$

$0.908 \times 800 = 726.4 \approx 726$

Hence, 726 homes will cost less than \$147,500.

(d) $z = \dfrac{\text{value} - \text{mean}}{\text{standard deviation}}$

$= \dfrac{139,000 - 145,500}{1500}$

$= -4.33$

The area between $z = 0$ and $z = -4.33$ is about 0.500. Since the desired area is to the right of -4.33, add 0.500 to 0.500.

$0.500 + 0.500 = 1.0$

$1.0 \times 800 = 800$

Hence, 800 homes will cost more than \$139,000.

19. Many real-life situations, with a large and random population, closely resemble the normal distribution (that is, the theoretical one). Now, the mathematics-statistics of this distribution are well known; hence, certain conclusions or probabilities can be drawn from an appropriate real-life situation.

21. Plot a graph, and see if the graph has properties quite similar to those of the normal distribution. (In more advanced treatments, certain tests do exist for deciding.)

23. Answers are approximate.

The As and Fs will have areas under the standard normal curve of $0.500 - 0.050 = 0.450$ and -0.450, respectively. Let $X_A =$ score that divides the As and $X_F =$ scre that divides the Fs. The z values corresponding to an area of 0.450 on either side of $z = 0$ are 1.64 and -1.64.

$$\frac{X_A - 60}{10} = 1.64; \quad X_A = 10(1.64) + 60 = 76.4$$

$$\frac{X_F - 60}{10} = -1.64; \quad X_F = 10(-1.64) + 60 = 43.6$$

The Cs will have an area under the standard normal curve of 0.300 on either side of $z = 0$. Let $X_{C1} =$ score that divides the Cs from Ds and let $X_{C2} =$ score that divides Cs from Bs. The z values corresponding to an area of 0.300 on either side of $z = 0$ are 0.84 and -0.84.

$$\frac{X_{C1} - 60}{10} = 0.84; \quad X_{C1} = 10(0.84) + 60 = 68.4$$

$$\frac{X_{C2} - 60}{10} = -0.84; \quad X_{C2} = 10(-0.84) + 60 = 51.6$$

The As will have scores above 76.
The Bs will have scores between 76 and 68.
The Cs will have scores between 68 and 52.
The Ds will have scores between 52 and 44.
The Fs will have scores of 44 and below.

25. The area under the normal curve between $z = 0$ and the cutoff time is $0.500 - 0.200 = 0.300$. The z value corresponding to an area of 0.300 to the left of $z = 0$ is -0.84.

$$\frac{\text{cutoff time} - 58.6}{4.3} = -0.84$$

$$\text{cutoff time} = 4.3(-0.84) + 58.6 = 55$$

The cutoff time is approximately 55 minutes.

Exercise Set 12-8

For this exercise set, use the following formulas.

$$r = \frac{n\left(\sum xy\right) - \left(\sum x\right)\left(\sum y\right)}{\sqrt{\left[n\left(\sum x^2\right) - \left(\sum x\right)^2\right]\left[n\left(\sum y^2\right) - \left(\sum y\right)^2\right]}}$$

$$b = \frac{n\left(\sum xy\right) - \left(\sum x\right)\left(\sum y\right)}{n\left(\sum x^2\right) - \left(\sum x\right)^2}$$

$$a = \frac{\sum y - b\left(\sum x\right)}{n}$$

1. (a)

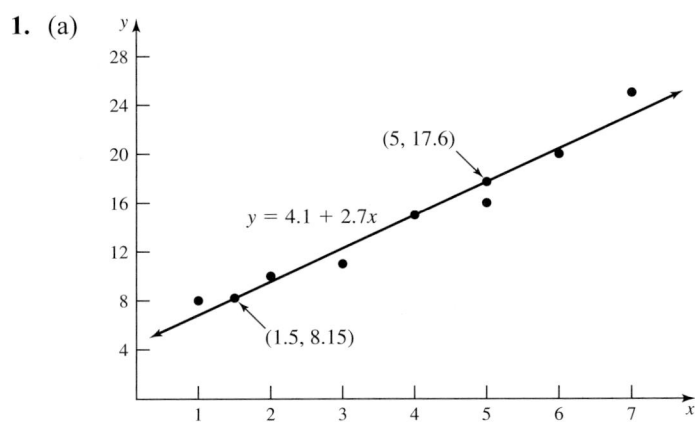

(b)

	x	y	xy	x^2	y^2
	1	8	8	1	64
	4	15	60	16	225
	6	20	120	36	400
	2	10	20	4	100
	3	11	33	9	121
	5	16	80	25	256
	7	25	175	49	625
Σ	28	105	496	140	1791

$$r = \frac{7(496)-28(105)}{\sqrt{[7(140)-(29)^2][7(1791)-(105)^2]}} = 0.977$$

(c) $n = 7$, 5% level, Appendix D value $= 0.754$
$n = 7$, 1% level, Appendix D value $= 0.875$
Since $|r| = 9.77$ is greater than each value, r is significant at the 5% and the 1% level.

(d) Since r is significant, draw the line. See graph in part (a).

$$b = \frac{7(496)-(28)(105)}{7(140)-(28)^2} = 2.714 = 2.7$$

$$a = \frac{105 - 2.714(28)}{7} = 4.1$$

The equation of the regression line is $y = 4.1 + 2.7x$.

(e) There is a positive linear relationship.

3. (a)

(b)

	x	y	xy	x^2	y^2
	75	10	750	5625	100
	80	5	400	6400	25
	85	11	935	7225	121
	90	4	360	8100	16
Σ	330	30	2445	27,350	262

$$r = \frac{4(2445) - 330(30)}{\sqrt{4(27,350) - (330)^2][4(262) - (30)^2]}} = -0.441$$

(c) $n = 4$, 5% level, Appendix D value = 0.950
$n = 4$, 1% level, Appendix D value = 0.990
Since $|r| = 0.441$ is not greater than either value, r is not significant at 5% nor at 1% level.

(d) Since r is not significant, the computing and drawing of a regression line would be meaningless.

(e) No relationship exists. The dots go "up and down."

5. (a)

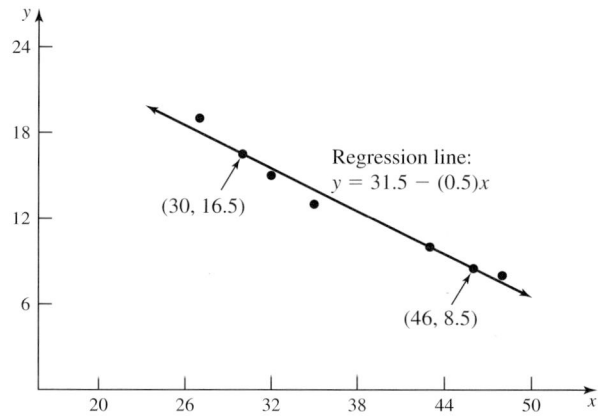

(b)

x	y	xy	x^2	y^2
27	19	513	729	361
35	13	455	1225	169
48	8	384	2304	64
43	10	430	1849	100
32	15	480	1024	225
Σ 185	65	2262	7131	919

$$\frac{5(2262) - (185)(65)}{\sqrt{[5(7131) - (185)^2][5(919) - (65)^2]}} = -0.983$$

(c) $n = 5$, 5% level, Appendix D value = 0.878
$n = 5$, 1% level, Appendix D value = 0.959
Since $|r| = 0.983$ is greater than each value, r is significant at the 5% and 1% level.

(d) Since r is significant, draw the line. See graph in part (a).
$$b = \frac{5(2262) - (185)(65)}{5(7131) - (185)^2} = -0.5$$
$$a = \frac{65 - (-0.5)(185)}{5} = 31.5$$

The equation of the regression line is $y = 31.5 - 0.5x$.

(e) There exists a negative linear relationship.

7. (a)

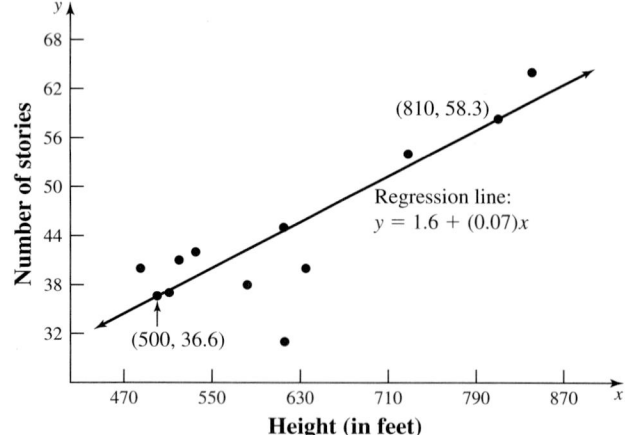

(b)

	x	y	xy	x^2	y^2
	485	40	19400	235225	1600
	511	37	18907	261121	1369
	520	41	21320	270400	1681
	535	42	22470	286225	1764
	582	38	22116	338724	1444
	615	45	27675	378225	2025
	616	31	19096	379456	961
	635	40	25400	403225	1600
	728	54	39312	529984	2916
	841	64	53824	707281	4096
Σ	6068	432	269,520	3,789,866	19,456

$$r = \frac{10(269,520) - (6068)(432)}{\sqrt{[10(3,789,866) - (6068)^2][10(19,456) - (432)^2]}} = 0.798$$

(c) $n = 10$, 5% level, Appendix D value $= 0.632$
$n = 10$, 1% level, Appendix D value $= 0.765$
Since $|r| = 0.798$ is larger than both values, r is significant at the 5% and 1% level.

(d) Since r is significant, draw the line. See graph in part (a).
$$b = \frac{10(269,520) - (6068)(432)}{10(3,789,866) - (6068)^2} = 0.06848 = 0.07$$
$$a = \frac{432 - (0.06848)(6068)}{10} = 1.6$$
The equation of the regression line is $y = 1.6 + 0.07x$.

(e) There is a positive linear relationship, except at the lower-left portion of the plot.
$y = 1.6 + 0.07(500) = 36.6$; when $x = 500$, y is predicted to be about 37.

9. (a)

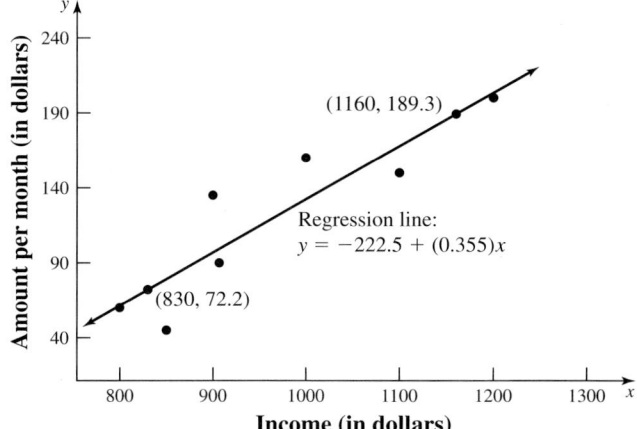

(b)

x	y	xy	x^2	y^2
800	60	48,000	640,000	3600
1200	200	240,000	1,440,000	40,000
1000	160	160,000	1,000,000	25,600
900	135	121,500	810,000	18,225
850	45	38,250	722,500	2025
907	90	81,630	822,649	8100
1100	150	165,000	1,210,000	22,500
Σ 6757	840	854,380	6,645,149	120,050

$$r = \frac{7(854,380) - (6757)(840)}{\sqrt{[7(6,645,149) - (6757)^2][7(120,050) - (840)^2]}} = 0.896$$

(c) $n = 7$, 5% level, Appendix D value $= 0.754$
$n = 7$, 1% level, Appendix D value $= 0.875$
Since $|r| = 0.896$ is larger than both values, r is significant at the 5% and 1% level.

(d) Since r is significant, draw the line. See graph in part (a).
$$b = \frac{7(854,380) - (6757)(840)}{7(6,645,149) - (6757)^2} = 0.3548 = 0.355$$
$$a = \frac{840 - (0.3548)(6757)}{7} = -222.5$$
The equation of the regression line is $y = -222.5 + 0.355x$.

(e) There is a positive linear relationship. $y = -222.5 + 0.355(925) = 105.875$; when $x = \$925$, y is predicted to be $\$105.88$.

11. (a)

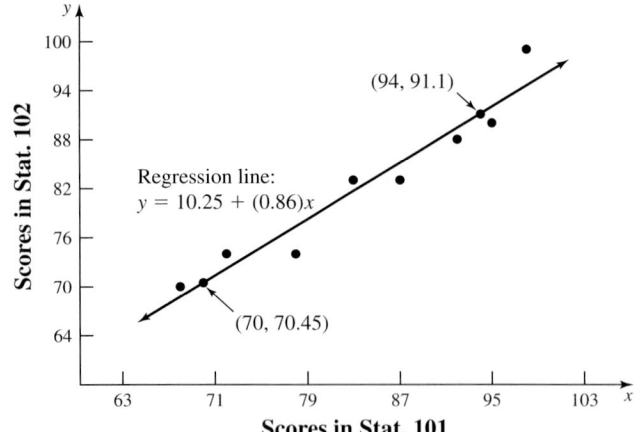

(b)

	x	y	xy	x^2	y^2
	87	83	7221	7569	6889
	92	88	8096	8464	7744
	68	70	4760	4624	4900
	72	74	5328	5184	5476
	95	90	8550	9025	8100
	78	74	5772	6084	5476
	83	83	6889	6889	6889
	98	99	9702	9604	9801
Σ	673	661	56,318	57,443	55,275

$$r = \frac{8(56,318) - (673)(661)}{\sqrt{[8(57,443) - (673)^2][8(55,275) - (661)^2]}} = 0.963$$

(c) $n = 8$, 5% level, Appendix D value = 0.707
$n = 8$, 1% level, Appendix D value = 0.834
Since $|r| = 0.963$ is larger than both values, r is significant at the 5% and 1% level.

(d) Since r is significant, draw the line. See graph in part (a).
$$b = \frac{8(56,318) - (673)(661)}{8(57,443) - (673)^2} = 0.8603 = 0.86$$
$$a = \frac{661 - (0.8603)(673)}{8} = 10.25$$
The equation of the regression line is $y = 10.25 + 0.86x$.

(e) There is a positive linear relationship.
$y = 10.25 + 0.86(90) = 87.65$; when $x = 90$, y is predicted to be 87.65.

13. Scatter plot

15. Generally, as x increases, so does y. The points would form a straight, or roughly straight, "stream" from lower left to upper right.

17. The values of r range from −1 to +1. At −1, a perfect negative linear relationship exists; at 0, no relationship exists; at +1, a perfect positive linear relationship exists.

19. Either +1 or −1

21.

x	y	xy	x^2	y^2
1	3	3	1	9
2	5	10	4	25
3	7	21	9	49
4	9	36	16	81
5	11	55	25	121
Σ 15	35	125	55	285

$$r = \frac{5(125) - (15)(35)}{\sqrt{[5(55) - (15)^2][5(285) - (35)^2]}} = 1$$

Interchange the values for x and y.

x	y	xy	x^2	y^2
3	1	3	9	1
5	2	10	25	4
7	3	21	49	9
9	4	36	81	16
11	5	55	121	25
Σ 35	15	125	285	55

$$r = \frac{5(125) - (35)(15)}{\sqrt{[5(285) - (35)^2][5(55) - (15)^2]}} = 1$$

The value of r is the same.

Review Exercises

1.

Item	Tally	Frequency
B	////	4
F	////	5
G	////	5
S	////	5
T	//// /	6

3.

Item	Cost ($) f	degrees $= \frac{f}{n} \cdot 360°$	percent $= \frac{f}{n} \cdot 100\%$
Food	4000	$\frac{4000}{16,000} \cdot 360° = 90°$	$\frac{4000}{16,000} \cdot 100\% = 25\%$
Clothing	1920	$\frac{1920}{16,000} \cdot 360° = 43.2°$	$\frac{1920}{16,000} \cdot 100\% = 12\%$
Savings	1600	$\frac{1600}{16,000} \cdot 360° = 36°$	$\frac{1600}{16,000} \cdot 100\% = 10\%$
Rent	4800	$\frac{4800}{16,000} \cdot 360° = 108°$	$\frac{4800}{16,000} \cdot 100\% = 30\%$
Other	3680	$\frac{3680}{16,000} \cdot 360° = 82.8°$	$\frac{3680}{16,000} \cdot 100\% = 23\%$

$n = 16,000$

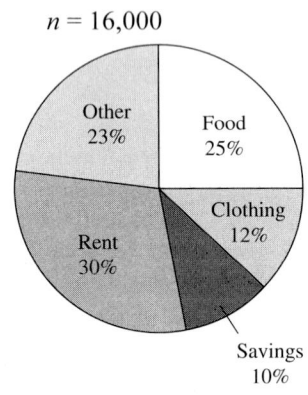

5. highest value − lowest value $= 187 − 102 = 85$

$$\frac{\text{difference}}{\text{number of classes}} = \frac{85}{6} \approx 14.2 \approx 15$$

Start with the lowest value and add 15 to get the lower class limits: 102, 117, 132, 147, 162, 177. Set up the classes by subtracting one from each lower class limit except the first lower class limit.

Rank	Tally	Frequency
102–116	////	4
117–131	///	3
132–146	/	1
147–161	////	4
162–176	̶//̶//̶/ ̶//̶//̶/ /	11
177–191	̶//̶//̶/ //	7

7. Represent the years on the x axis and the amount on the y axis, and then draw straight lines through the points.

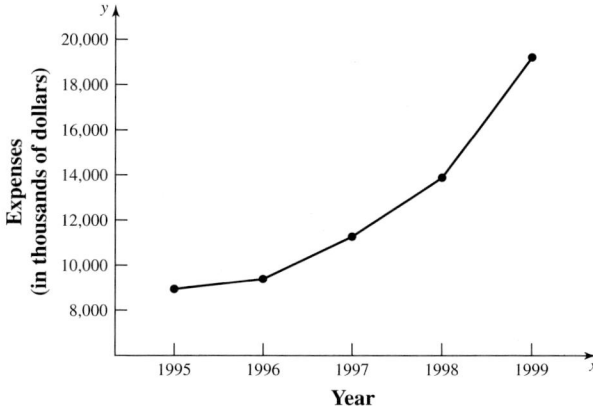

9.

Hours	Frequency	Midpoint	Frequency × Midpoint
1–3	1	2	2
4–6	4	5	20
7–9	5	8	40
10–12	1	11	11
13–15	1	14	14
	12		87

$$\overline{X} = \frac{87}{12} = 7.25$$

11. Arrange the data in order.
29 45 74 91 118 151 158 285
The median is the middle point.

$$Q_2 = \frac{91+118}{2} = 104.5$$

Find the median of the values less than Q_2.

$$Q_1 = \frac{45+74}{2} = 59.5$$

Find the median of the values above Q_2.

$$Q_3 = \frac{151+158}{2} = 154.5$$

13. (a) $z = \dfrac{\text{value} - \text{mean}}{\text{standard deviation}} = \dfrac{4-3}{\frac{1}{3}} = 3$

the area between $z = 0$ and $z = 3$ is 0.499. Since the desired area is in the right tail, subtract 0.499 from 0.500.
0.500 − 0.499 = 0.001
Hence, the probability that it will take more than 4 years is 0.001, or 0.1%.

(b) $z = \dfrac{\text{value} - \text{mean}}{\text{standard deviation}} = \dfrac{3-3}{\frac{1}{3}} = 0$

The area less than $z = 0$ is 0.500. Hence, the probability that it will take less than 3 years in 0.5, or 50%.

(c) $z_1 = \dfrac{\text{value} - \text{mean}}{\text{standard deviation}} = \dfrac{3.8 - 3}{\frac{1}{3}} = 2.4$

$z_2 = \dfrac{\text{value} - \text{mean}}{\text{standard deviation}} = \dfrac{4.5 - 3}{\frac{1}{3}} = 4.5$

The area between $z = 0$ and $z = 2.4$ is 0.492. The area between $z = 0$ and $z = 4.5$ is about 0.500. The area between $z = 2.4$ and $z = 4.5$ is $0.500 - 0.492 = 0.008$. Hence, the probability that it will take between 3.8 and 4.5 years is 0.008, or 0.8%

(d) $z_1 = \dfrac{\text{value} - \text{mean}}{\text{standard deviation}} = \dfrac{2.5 - 3}{\frac{1}{3}} = -1.5$

$z_2 = \dfrac{\text{value} - \text{mean}}{\text{standard deviation}} = \dfrac{3.1 - 3}{\frac{1}{3}} = 0.3$

The area between $z = 0$ and $z = -1.5$ is 0.433. The area between $z = 0$ and $z = 0.3$ is 0.118. The area between $z = -1.5$ and $z = 0.3$ is $0.433 + 0.118 = 0.551$. Hence, the probability that it will take between 2.5 and 3.1 years is 0.551, or 55.1%.

15. $z = \dfrac{\text{value} - \text{mean}}{\text{standard deviation}} = \dfrac{43.5 - 45}{2} = -0.75$

The are between $z = 0$ and $z = -0.75$ is 0.273. Since the desired area is in the left tail subtract 0.273 from 0.500.
$0.500 - 0.273 = 0.227$
$0.277 \times 2000 = 454$
Hence, 454 will weigh less than 43.5 pounds.

17.

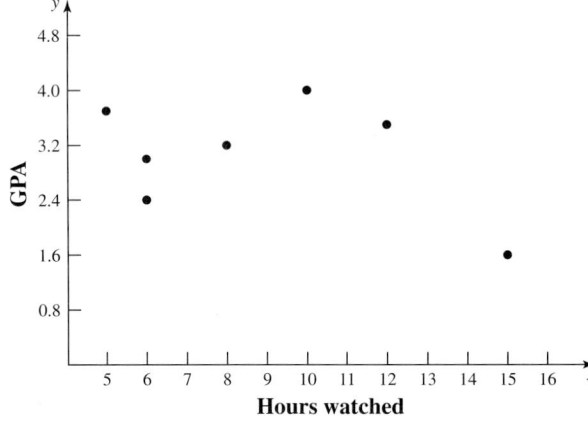

There is no discernable relationship.

	x	y	xy	x^2	y^2
	6	2.4	14.4	36	5.76
	10	4.0	40.0	100	16.00
	8	3.2	25.6	64	10.24
	15	1.6	24.0	225	2.56
	5	3.7	18.5	25	13.69
	6	3.0	18.0	36	9.00
	12	3.5	42.0	144	12.25
Σ	62	21.4	182.5	630	69.5

$$r = \frac{7(182.5) - (62)(21.4)}{\sqrt{[7(630) - (62)^2][7(69.5) - (21.4)^2]}} = -0.388$$

For $n = 7$ and 5% level, the value in Appendix D is 0.754. Since $|r| = 0.388$ is not larger than this value, r is not significant at the 5% level.

No regression line is appropriate (since r is not significant), and no prediction for $x = 9$ is appropriate.

Chapter Test

1.

Source	Tally	Frequency
M	༒ ⁄	6
N	༒ ⁄⁄	7
R	༒ ⁄⁄	7
T	༒	5

3.

Source	Frequency f	degrees $= \frac{f}{n} \cdot 360°$	percent $= \frac{f}{n} \cdot 100\%$
M	6	$\frac{6}{25} \cdot 360° = 86.4°$	$\frac{6}{25} \cdot 100\% = 24\%$
N	7	$\frac{7}{25} \cdot 360° = 100.8°$	$\frac{7}{25} \cdot 100\% = 28\%$
R	7	$\frac{7}{25} \cdot 360° = 100.8°$	$\frac{7}{25} \cdot 100\% = 28\%$
T	5	$\frac{5}{25} \cdot 360° = 72°$	$\frac{5}{25} \cdot 100\% = 20\%$

$n = 25$

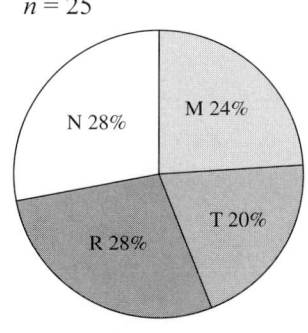

5. For the histogram, draw vertical bars corresponding to the frequencies for each class.

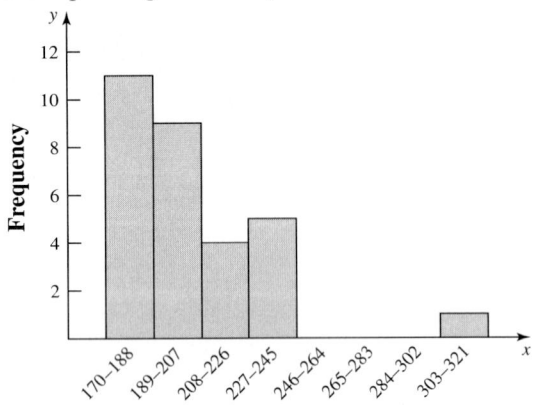

For the frequency polygon, find the midpoints for each class: 179, 198, 217, 236, 255, 274, 293, and 312. Label the horizontal axis with the midpoints. Connect adjacent midpoints with straight lines. Finish the graph by drawing a line back to the horizontal at the beginning and end of graph.

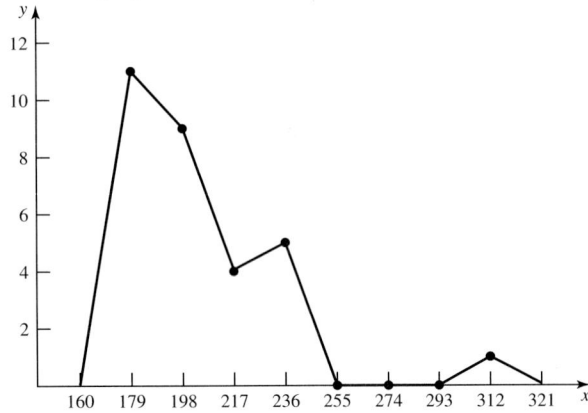

7. Represent the years on the x axis and the wages on the y axis, and then draw straight lines through the points.

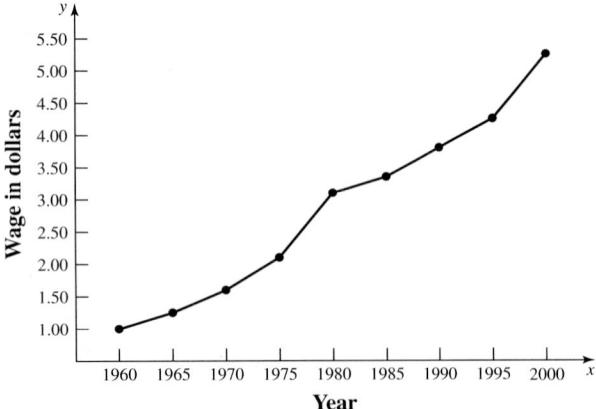

The graph shows an increase in the wage for all periods. The two steepest increases were during the jumps from 1975 to 1980 and from 1995 to 2000.

9.

Errors	Frequency	Midpoint	Frequency × Midpoint
0–2	1	1	1
3–5	3	4	12
6–8	4	7	28
9–11	1	10	10
12–14	1	13	13
	10		64

$$\bar{X} = \frac{64}{10} = 6.4$$

11. (a)

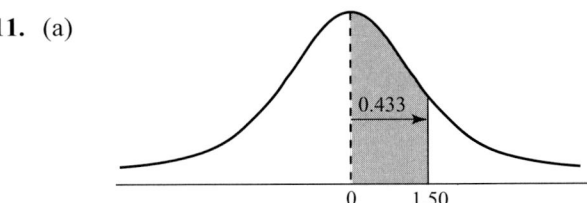

The area between $z = 0$ and $z = 1.50$ is 0.433.

(b)

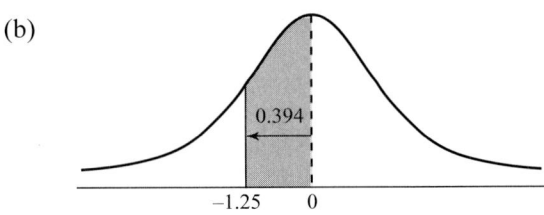

The area between $z = 0$ and $z = -1.25$ is 0.394.

(c)

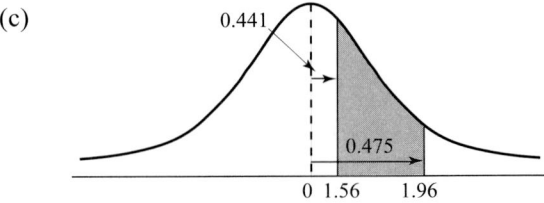

The area between $z = 1.56$ and $z = 1.96$ is $0.475 - 0.441 = 0.034$.

(d)

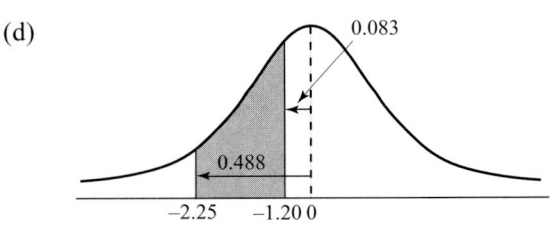

The area between $z = -1.20$ and $z = -2.25$ is $0.488 - 0.385 = 0.103$.

(e)

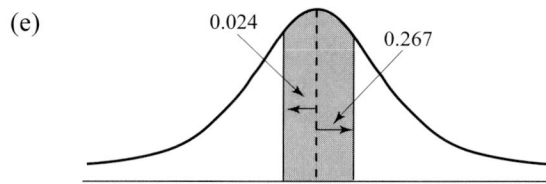

The area between $z = -0.06$ and $z = 0.73$ is $0.024 + 0.267 = 0.291$.

(f)

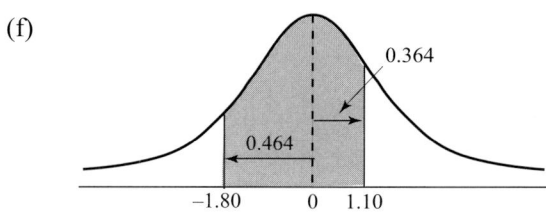

The area between $z = 1.10$ and $z = -1.80$ is $0.464 + 0.364 = 0.828$.

(g)

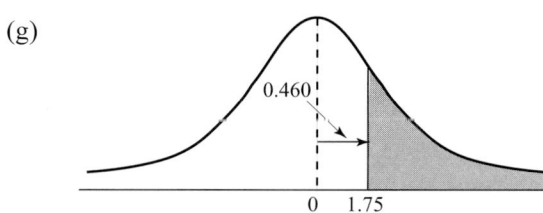

The area to the right of $z = 1.75$ is $0.500 - 0.460 = 0.040$.

(h)

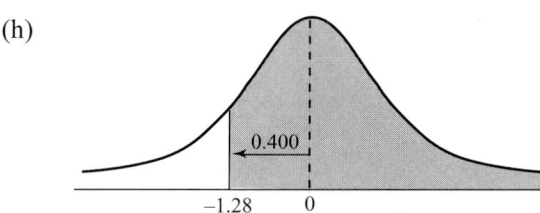

The area to the right of $z = -1.28$ is $0.500 + 0.400 = 0.900$.

(i)

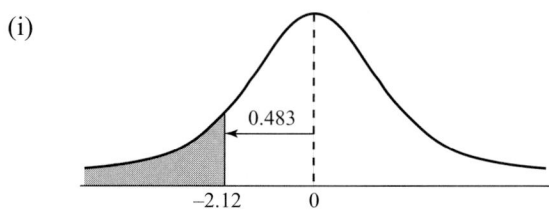

The area to the left of $z = -2.12$ is $0.500 - 0.483 = 0.017$.

(j)

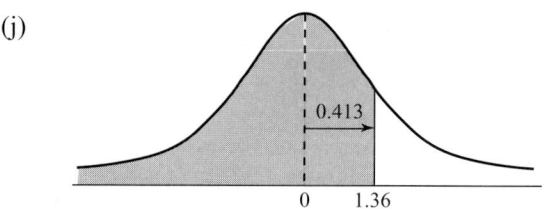

The area to the left of $z = 1.36$ is $0.500 + 0.413 = 0.913$.

13. (a) $z = \dfrac{\text{value} - \text{mean}}{\text{standard deviation}} = \dfrac{59 - 53}{4} = 1.5$

The area between $z = 0$ and $z = 1.5$ and 0.433. Since the desired area is in the right tail, subtract 0.433 from 0.500.
$0.500 - 0.433 = 0.067$
Hence, the probability that the height is greater than 59 inches is 0.067, or 6.7%.

(b) $z = \dfrac{\text{value} - \text{mean}}{\text{standard deviation}} = \dfrac{45 - 53}{4} = -2$

The area between $z = 0$ and $z = -2$ is 0.477. Since the desired area is in the left tail, subtract 0.477 from 0.500.
$0.500 - 0.477 = 0.023$
Hence, the probability that the height is less than 45 inches is 0.023, or 2.3%.

(c) $z_1 = \dfrac{\text{value} - \text{mean}}{\text{standard deviation}} = \dfrac{50 - 53}{4} = -0.75$

$z_2 = \dfrac{\text{value} - \text{mean}}{\text{standard deviation}} = \dfrac{55 - 53}{4} = 0.5$

The area between $z = 0$ and $z = -0.75$ is 0.273. The area between $z = 0$ and $z = 0.5$ is 0.192. The desired area is $0.273 + 0.192 = 0.465$. Hence, the probability that the height is between 50 and 55 inches is 0.465, or 46.5%.

(d) $z_1 = \dfrac{\text{value} - \text{mean}}{\text{standard deviation}} = \dfrac{58 - 53}{4} = 1.25$

$z_2 = \dfrac{\text{value} - \text{mean}}{\text{standard deviation}} = \dfrac{62 - 53}{4} = 2.25$

The area between $z = 0$ and $z = 1.25$ is 0.394. The area between $z = 0$ and $z = 2.25$ is 0.488. The desired area is $0.488 - 0.394 = 0.094$.
Hence, the probability that the height is between 58 and 62 inches is 0.094, or 9.4%.

15.

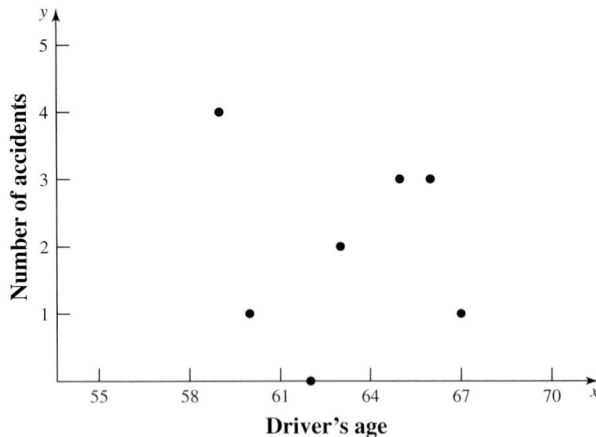

No relationship can be discerned from the scatter plot.

x	y	xy	x^2	y^2
63	2	126	3969	4
65	3	195	4225	9
60	1	60	3600	1
62	0	0	3844	0
66	3	198	4356	9
67	1	67	4489	1
59	4	236	3481	16
Σ 442	14	882	27,964	40

$$r = \frac{7(882) - (442)(14)}{\sqrt{[7(27,964) - (442)^2][7(40) - (14)^2]}} = -0.078$$

For $n = 7$ and 5% level, the value in Appendix D is 0.754. Since $|r| = 0.078$ is not larger than the value, r is not significant at the 5% level. Hence, no regression equation and no prediction are appropriate.

Chapter Supplement

1. (a) The number of people (20) is too small of a sample when compared to the general population of, say, the state of Minnesota.

   (b) How were the 20 subjects chosen? Did they represent the general population, or were they chosen from a specific group such as college students or others?

3. Various questions need to be answered:

   (a) "Less traveled" in what respect?

   (b) How many groups of 100 women were there?

   (c) Were the women selected from the general population or from a specific group?

5. Perhaps the article's originator wanted to deceive readers into associating 11 with 18 (thus exceeding more than half), whereas 11% of 18 is only $1.98 \approx 2$. Moreover, the article should address the presence or absence of side effects.

7. The ad does not state what the "74% more" is compared with, nor how the conclusion is reached.

9. The words "acid control" are vague. Is the control, say, 5%, 50%, or what? Does the word "control" mean neutralization of acid, and into what? Or, does it mean nonproduction of acid? If Brand X is indeed effective, are there side effects?

11. The two graphs do not have any labeling or any scales on the vertical axes. Hence, the apparent faster speed in the first graph is not necessarily appreciable, as the graph "suggests"; in fact, the two speeds might be very close.

13. In the second graph, the vertical distance to represent $1 is much bigger than in the first graph. Accordingly, "steepness" and "jumps" are much stronger in the second graph.

15. Disadvantages to the safety locks should be pointed out, such as: additional cost, additional time to operate the gun, etc.

17. Responders may not be telling the truth.

19. Better methods of detecting lead poisoning have been developed.

21. There are no factors given to make comparisons.

13 Voting Methods

Exercise Set 13-1

1.

Number of votes	5	4	8	5
First choice	X	X	Y	Z
Second choice	Y	Z	Z	Y
Third choice	Z	Y	X	X

(a) Find the sum of the numbers in the top row.
$5 + 4 + 8 + 5 = 22$

(b) Look at the number above column 2. Hence, 4 people voted in the order XZY.

(c) Only column 3 has Y as the first choice. Hence, 8 people voted for Y as their first choice.

(d) Consider only the first-place votes.
X received $5 + 4$ or 9
Y received 8
Z received 5
Hence, candidate X is the winner.

3.

Number of votes	9	4	5
First choice	P	C	M
Second choice	M	P	P
Third choice	C	M	C

(a) Find the sum of the numbers in the top row.
$9 + 4 + 5 = 18$

(b) Look at the number in column 1. Hence, 9 people voted in the order PMC.

(c) Only column 2 has C as the first choice. Hence, 4 people voted for Chicago as their first choice.

(d) Consider only the first-place votes.
P received 9
C received 4
M received 5
Hence, Philadelphia is the winner.

5. (a) Find the sum of the numbers in the top row.
$83 + 56 + 42 + 27 = 208$

(b) Consider only the first-place votes.
G received 83
S received $56 + 42$ or 98
B received 27
Hence, swimming pool is the winner.

7. (a) Find the sum of the numbers in the top row.
$3 + 5 + 2 + 6 + 4 = 20$

(b) Consider only the first-place votes.
R received 3 + 4 or 7
D received 5
C received 2 + 6 or 8
Hence, carnations is the winner.

9.

Column	—	was preferred over	—	voters	head-to-head
1	G		S	83	
2	S		G	56	
3	S		G	42	S over G
4	S		G	27	
1	S		B	83	
2	S		B	56	
3	S		B	42	S over B
4	B		S	27	
1	G		B	83	
2	G		B	56	
3	B		G	42	G over B
4	B		G	27	

The criterion has not been violated because in a head-to-head comparison, S won over G and B.

11.

Column	— was preferred over —		voters	head-to-head
1	R	G	3	
2	G	R	5	
3	R	G	2	G over R
4	G	R	6	
5	R	G	4	
1	R	C	3	
2	C	R	5	
3	C	R	2	C over R
4	C	R	6	
5	R	C	4	
1	R	D	3	
2	D	R	5	
3	R	D	2	D over R
4	D	R	6	
5	R	D	4	
1	G	C	3	
2	C	G	5	
3	C	G	2	C over G
4	C	G	6	
5	C	G	4	
1	G	D	3	
2	D	G	5	
3	G	D	2	G over D
4	G	D	6	
5	D	G	4	
1	C	D	3	
2	D	C	5	
3	C	D	2	C over D
4	C	D	6	
5	C	D	4	

The criterion has not been violated because in a head-to-head comparison, C won over R, G, and D.

13. A preference table is a summary of an election where candidates are ranked by voters as to first choice, second choice, etc.

15. The head-to-head comparison criterion states that if a particular candidate wins all head-to-head comparisons with all other candidates, then the candidate should win the election.

17. To assure winning the election, candidate A needs a total of 51 votes. Hence, to be assured of winning the election, candidate A needs 51 – 36 or 15 votes. However, candidate A can also win if all the remaining 20 votes go to candidate C since C would have a total of 32 votes. Other possibilities also exist.

19. No; if candidate C received all of the 20 remaining votes, he or she would only have a total of 32 votes and candidate A would win with 36 votes.

21. Yes; two candidates could have tied in the number of voters that preferred them.

Exercise Set 13-2

1. Type
S: $8 \cdot 3 + 5 \cdot 2 + 4 \cdot 1 + 2 \cdot 1 = 40$
B: $8 \cdot 2 + 5 \cdot 1 + 4 \cdot 2 + 2 \cdot 3 = 35$
N: $8 \cdot 1 + 5 \cdot 3 + 4 \cdot 3 + 2 \cdot 2 = 39$
The winner is science fiction (S).

3. Movie
G: $331 \cdot 4 + 317 \cdot 1 + 206 \cdot 2 + 98 \cdot 3 = 2347$
A: $331 \cdot 3 + 317 \cdot 4 + 206 \cdot 1 + 98 \cdot 2 = 2663$
C: $331 \cdot 2 + 317 \cdot 3 + 206 \cdot 4 + 98 \cdot 1 = 2535$
B: $331 \cdot 1 + 317 \cdot 2 + 206 \cdot 3 + 98 \cdot 4 = 1975$
The winner is *Anatomy of a Murder* (A).

5. (a) Improvement
G: $83 \cdot 3 + 56 \cdot 2 + 42 \cdot 1 + 27 \cdot 1 = 430$
S: $83 \cdot 2 + 56 \cdot 3 + 42 \cdot 3 + 27 \cdot 2 = 514$
B: $83 \cdot 1 + 56 \cdot 1 + 42 \cdot 2 + 27 \cdot 3 = 304$
The winner is build a swimming pool (S).

(b) Yes; the winner is the same as the one determined by the plurality method.

7. Yes, since no book type received the majority of first-place votes.
$\left(\text{Note: majority is} \geq 10; \dfrac{8+5+4+2}{2} = \dfrac{19}{2} = 9.5.\right)$

9. Yes, since no movie received the majority of first-place votes.
$\left(\text{Note: majority is} \geq 477; \dfrac{331+317+206+98}{2} = \dfrac{952}{2} = 476.\right)$

11. Yes, since no choice received the majority of first-place votes.
$\left(\text{Note: majority is} \geq 105; \dfrac{83+56+42+27}{2} = \dfrac{208}{2} = 104.\right)$

13. majority ≥ 5; $\dfrac{4+3+2}{2} = \dfrac{9}{2} = 4.5$

No one received a majority of votes, so candidate G is eliminated since he or she has the fewest first-place votes. New preference table:

Number of votes	4	3	2
First choice	D	W	D
Second choice	W	D	W

The winner is Professor Donovan (D) with 6 (4 + 2) first-place votes.

15. (a) majority ≥ 11; $\dfrac{3+5+2+6+4}{2} = \dfrac{20}{2} = 10$

None received a majority of votes, so choice D is eliminated since D received the fewest first-place votes. New preference table:

Number of votes	3	5	2	6	4
First choice	R	C	C	C	R
Second choice	G	G	R	G	C
Third choice	C	R	G	R	G

The winner is carnations (C) with 12 (2 + 6 + 4) first-place votes.

(b) The winner is the same when using the plurality method.

17. No; the election does not violate the monotonicity criterion since Professor Donovan is still the winner.

19. No; the election does not violate the monotonicity criterion since carnations is still the winner.

21. The Borda count method of voting requires that each candidate be ranked from most favorable to least favorable, and then 1 point is assigned to the last-place candidate, 2 points to the next-to-last place candidate, etc. The candidate with the most points is declared the winner.

23. The plurality-with-elimination method states that if a candidate has a majority of first-place votes, that candidate is declared the winner. If no candidate has a majority of first-place votes, the candidate with the least number of first-place votes is eliminated, and then another count is taken with each of the other candidates moving up. This continues until a candidate receives a majority of first-place votes.

25. Answers will vary.

27. Answers will vary.

29. Yes; now the winner will have the lowest total number of ranked points.

Exercise Set 13-3

1. $\dfrac{n(n-1)}{2} = \dfrac{4(4-1)}{2} = \dfrac{4(3)}{2} = 6$

3. $\dfrac{n(n-1)}{2} = \dfrac{10(10-1)}{2} = \dfrac{10(9)}{2} = 45$

5.

choice	votes				winner
	26	19	15	6	
1	Ⓡ	T	Ⓢ	Ⓡ	
2	Ⓢ	Ⓢ	Ⓡ	T	S
3	T	Ⓡ	T	Ⓢ	
1	Ⓡ	Ⓣ	S	Ⓡ	
2	S	S	Ⓡ	Ⓣ	R
3	Ⓣ	Ⓡ	Ⓣ	S	
1	R	Ⓣ	Ⓢ	R	
2	Ⓢ	Ⓢ	R	Ⓣ	S
3	Ⓣ	R	Ⓣ	Ⓢ	

Point totals
R   1
S   1 + 1 = 2
T   0
Hence, Steel Center (S) is the winner.

7. (a)

choice	votes			winner
	4	3	2	
1	D	Ⓦ	Ⓖ	
2	Ⓦ	Ⓖ	D	W
3	Ⓖ	D	Ⓦ	
1	Ⓓ	W	Ⓖ	
2	W	Ⓖ	Ⓓ	G
3	Ⓖ	Ⓓ	W	
1	Ⓓ	Ⓦ	G	
2	Ⓦ	G	Ⓓ	D
3	G	Ⓓ	Ⓦ	

Point Totals
G   1
W   1
D   1
There is a three-way tie.

(b) The results are different since Professor Donovan won when using the plurality-with-elimination method.

9. (a)

choice	votes 8	6	5	2	winner
1	R	W	(C)	(F)	
2	W	R	(F)	R	C
3	(C)	(F)	R	(C)	
4	(F)	(C)	W	W	
1	R	(W)	(C)	F	
2	(W)	R	F	F	W
3	(C)	F	R	(C)	
4	F	(C)	(W)	(W)	
1	(R)	W	(C)	F	
2	W	(R)	F	(R)	R
3	(C)	F	(R)	(C)	
4	F	(C)	W	W	
1	R	(W)	C	(F)	
2	(W)	R	(F)	R	W
3	C	(F)	R	C	
4	(F)	C	(W)	(W)	
1	(R)	W	C	(F)	
2	W	(R)	(F)	(R)	R
3	C	(F)	(R)	C	
4	(F)	C	W	W	
1	(R)	(W)	C	F	
2	(W)	(R)	F	(R)	R
3	C	F	(R)	C	
4	F	C	(W)	(W)	

Point totals
C   1
F   0
W   1 + 1 = 2
R   1 + 1 + 1 = 3
Hence, Rosa's Restaurant (R) is the winner.

(b) Yes; Rosa's Restaurant also won when the Borda count method was used.

11. Preference table with D removed.

Number of votes	4	3	2
First choice	W	W	G
Second choice	G	G	W

Yes, this election violates the irrelevant alternative criterion since Professor Williams (W) wins.

13. Preference table with W removed.

Number of votes	8	6	5	2
First choice	R	R	C	F
Second choice	C	F	F	R
Third choice	F	C	R	C

No, this election does not violate the irrelevant alternative criterion since Rosa's Restaurant is still the winner.

15.

Candidate	Votes
Dr. Michaels	42
Dr. Jones	45
Dr. Philip	43
Dr. Smith	43

The winner is Dr. Jones.

17.

Color	Votes
White	7
Blue	4
Green	8
Silver	5

The winner is green.

19.

Inmate	Votes
W	4
X	3
Y	3
Z	5

The winner is inmate Z.

21. Each candidate is ranked by the voters. Then each candidate is paired with every other candidate in a head-to-head contest. The winner of each contest gets 1 point. In case of a tie, each candidate gets a $\frac{1}{2}$ point. The candidate with the most points wins the election.

23. With approval voting, each voter gives one vote to as many candidates on the ballot as he or she finds acceptable. The votes are counted, and the winner is the candidate who receives the most votes.

25. Answers will vary.

27. Answers will vary.

Review Exercises

1.

Number of votes	5	5	5
First choice	Q	P	R
Second choice	R	Q	P
Third choice	P	R	Q

3. Only column 3 has R as the first choice. Hence, 5 people voted for Ross as the best speaker.

5.

Number of votes	6	4	10
First choice	C	P	H
Second choice	P	C	P
Third choice	H	H	C

7. Only column 1 has C as the first choice. Hence, 6 people voted for Pizza City as their first choice.

9. Find the sum of the numbers in the top row.
$26 + 15 + 10 + 7 = 58$

11. Consider only the first-place votes.
A received 26
B received $15 + 7 = 22$
C received 10
Hence, style A won.

13. majority ≥ 30; $\dfrac{26+15+10+7}{2} = \dfrac{58}{2} = 29$

None received a majority of votes, so choice C is eliminated since it received the fewest first place votes.
New preference table:

Number of votes	26	15	10	7
First choice	A	B	A	B
Second choice	B	A	B	A

The winner is style A with 36 (26 + 10) first-place votes.

15.

Column	—	was preferred over	—	voters	head-to-head
1	A		B	26	
2	B		A	15	
3	A		B	10	A over B
4	B		A	7	
1	A		C	26	
2	C		A	15	
3	C		A	10	A over C
4	A		C	7	
1	B		C	26	
2	B		C	15	
3	C		B	10	B over C
4	B		C	7	

No; the criterion has not been violated because in a head-to-head comparison, A won over B and C.

17. Preference table with C removed.

Number of votes	26	15	10	7
First choice	A	B	A	B
Second choice	B	A	B	A

No, this election does not violate the irrelevant alternative criterion since style A is still the winner.

19. Find the sum of the numbers in the top row.
$18 + 17 + 9 + 3 = 47$

21. Consider only the first-place votes.
C received 3
G received 9
B received 17
M received 18
Hence, *Music Man* (M) won.

23. majority ≥ 24; $\dfrac{18+17+9+3}{2} = \dfrac{47}{2} = 23.5$

None received a majority of votes, so C is eliminated since it received the fewest first-place votes. New preference table:

Number of votes	18	17	9	3
First choice	M	B	G	G
Second choice	B	G	M	B
Third choice	G	M	B	M

Still none received a majority of votes, so choice G is eliminated since it received the fewest first-place votes. New preference table:

Number of votes	18	17	9	3
First choice	M	B	M	B
Second choice	B	M	B	M

The winner is *Music Man* (M) with 27 (18 + 9) first-place votes.

25.

Column	— was preferred over	—	voters	head-to-head
1	G	C	18	
2	G	C	17	
3	G	C	9	G over C
4	C	G	3	
1	B	C	18	
2	B	C	17	
3	C	B	9	B over C
4	C	B	3	
1	M	C	18	
2	C	M	17	
3	C	M	9	C over M
4	C	M	3	
1	B	G	18	
2	B	G	17	
3	G	B	9	B over G
4	G	B	3	
1	M	G	18	
2	G	M	17	
3	G	M	9	G over M
4	G	M	3	
1	M	B	18	
2	B	M	17	
3	M	B	9	M over B
4	B	M	3	

Yes; the criterion has been violated because in a head-to-head comparison, M won over only B.

27. No; the monotonicity criterion was not violated.

29.

Program	Votes
Magician	5
Speaker	9
Rock band	9
Comedian	5

There is a tie between a speaker and a rock band.

Chapter Test

1.

Number of votes	1	7	4
First choice	A	B	C
Second choice	B	A	B
Third choice	C	C	A

3. Look at the votes in column 2 where B is the first choice. Hence, 7 people voted for brand B as their first choice.

5. Find the sum of the numbers in the top row.
$43 + 27 + 18 + 12 = 100$

7. Consider only the first-place votes.
P received $43 + 12 = 55$
R received 27
B received 18
Hence, Pittsburgh (P) won.

9. Majority ≥ 51; $\dfrac{43 + 27 + 18 + 12}{2} = 50$

None received a majority of votes, so city B is eliminated since it received the fewest first-place votes. New preference table:

Number of votes	43	27	18	12
First choice	P	R	R	P
Second choice	R	P	P	R

The winner is Pittsburgh (P) with 55 (43 + 12) first-place votes.

11.

Column	— was preferred over —		voters	head-to-head
1	P	B	43	
2	P	B	27	
3	B	P	18	P over B
4	P	B	12	
1	P	R	43	
2	R	P	27	
3	R	P	18	P over R
4	P	R	12	
1	R	B	43	
2	R	B	27	
3	B	R	18	R over B
4	B	R	12	

No; the criterion has not been violated because in a head-to-head comparison, P won over B and R.

13. No, Pittsburgh (P) is the winner.
See Exercises 9 and 11.

15.

Physician	Votes
Dr. Michaels	42
Dr. Jones	45
Dr. Philip	43
Dr. Spoz	43

The winner is Dr. Jones.

Appendix A | Measurement

Exercise Set A-1

1. There are 2 steps from meters to centimeters.
$8 \text{ m} \times 10^2 = 8 \times 100 = 800 \text{ cm}$

3. There is 1 step from dekameters to meters.
$12 \text{ dam} \times 10^1 = 12 \times 10 = 120 \text{ m}$

5. There is 1 step from kilometers to hectometers.
$0.6 \text{ km} \times 10^1 = 0.6 \times 10 = 6 \text{ hm}$

7. There is 1 step from meters to decimeters.
$90 \text{ m} \times 10^1 = 90 \times 10 = 900 \text{ dm}$

9. There are 2 steps from centimeters to meters.
$375.6 \text{ cm} \div 10^2 = 376.6 \div 100 = 3.756 \text{ m}$

11. There are 3 steps from meters to kilometers.
$405.3 \text{ m} \div 10^3 = 405.3 \div 1000 = 0.4053 \text{ km}$

13. There are 5 steps from kilometers to centimeters.
$12 \text{ km} \times 10^5 = 12 \times 100,000 = 1,200,000 \text{ cm}$

15. There are 6 steps from kilometers to millimeters.
$1.85 \text{ km} \times 10^6 = 1.85 \times 1,000,000$
$= 1,850,000 \text{ mm}$

17. There are 4 steps from kilometers to decimeters.
$12.62 \text{ km} \times 10^4 = 12.62 \times 10,000 = 126,200 \text{ dm}$

19. There is 1 step from hectometers to kilometers.
$8 \text{ hm} \div 10^1 = 8 \div 10 = 0.8 \text{ km}$

21. There are 3 steps from milliliters to liters.
$500 \text{ mL} \div 10^3 = 500 \div 1000 = 0.5 \text{ L}$

23. There is 1 step from liters to deciliters.
$92 \text{ L} \times 10^1 = 92 \times 10 = 920 \text{ dL}$

25. There is 1 step from deciliters to centiliters.
$8 \text{ dL} \times 10^1 = 8 \times 10 = 80 \text{ cL}$

27. There is 1 step from hectoliters to dekaliters.
$6.7 \text{ hL} \times 10^1 = 6.7 \times 10 = 67 \text{ daL}$

29. There is 1 step from kiloliters to hectoliters.
$64 \text{ kL} \times 10^1 = 64 \times 10 = 640 \text{ hL}$

31. There are 3 steps from liters to kiloliters.
$81 \text{ L} \div 10^3 = 81 \div 1000 = 0.081 \text{ kL}$

33. There is 1 step from liters to dekaliters.
$7 \text{ L} \div 10^1 = 7 \div 10 = 0.7 \text{ daL}$

35. There is 1 step from liters to dekaliters.
$117 \text{ L} \div 10^1 = 117 \div 10 = 11.7 \text{ daL}$

37. There are 2 steps from centiliters to liters.
$142 \text{ cL} \div 10^2 = 142 \div 100 = 1.42 \text{ L}$

39. There are 6 steps from milliliters to kiloliters.
$32,546 \text{ mL} \div 10^6 = 32,546 \div 1,000,000$
$= 0.032546 \text{ kL}$

41. There is 1 step from decigrams to centigrams.
$9 \text{ dg} \times 10^1 = 9 \times 10 = 90 \text{ cg}$

43. There is 1 step from kilograms to hectograms.
$11 \text{ kg} \times 10^1 = 11 \times 10 = 110 \text{ hg}$

45. There is 1 step from decigrams to centigrams.
$18 \text{ dg} \times 10^1 = 18 \times 10 = 180 \text{ cg}$

47. There is 1 step from centigrams to decigrams.
$71 \text{ cg} \div 10^1 = 71 \div 10 = 7.1 \text{ dg}$

49. There is 1 step from grams to decigrams.
$5 \text{ g} \times 10^1 = 5 \times 10 = 50 \text{ dg}$

51. There are 3 steps from grams to milligrams.
$0.325 \text{ g} \times 10^3 = 0.325 \times 1000 = 325 \text{ mg}$

53. There are 3 steps from kilograms to grams.
$4325 \text{ kg} \times 10^3 = 4325 \times 1000 = 4,325,000 \text{ g}$

55. There are 3 steps from milligrams to grams.
$86 \text{ mg} \div 10^3 = 86 \div 1000 = 0.086 \text{ g}$

57. There are 3 steps from grams to kilograms.
$400 \text{ g} \div 10^3 = 400 \div 1000 = 0.4 \text{ kg}$

59. There are 2 steps from grams to hectograms.
$5632 \text{ g} \div 10^2 = 5632 \div 100 = 56.32 \text{ hg}$

Exercise Set A-2

1. $5 \, \cancel{\text{m}} \cdot \dfrac{100 \, \cancel{\text{cm}}}{1 \, \cancel{\text{m}}} \cdot \dfrac{1 \text{ in.}}{2.54 \, \cancel{\text{cm}}} = \dfrac{5 \cdot 100 \text{ in.}}{2.54} \approx 196.85 \text{ in.}$

3. $16 \text{ in.} \cdot \dfrac{2.54 \text{ cm}}{1 \text{ in.}} \cdot \dfrac{10 \text{ mm}}{1 \text{ cm}} = \dfrac{16 \cdot 2.54 \cdot 10 \text{ mm}}{1}$
$= 406.4 \text{ mm}$

5. $235 \text{ ft} \cdot \dfrac{30.48 \text{ cm}}{1 \text{ ft}} \cdot \dfrac{1 \text{ dam}}{1000 \text{ cm}} = \dfrac{235 \cdot 30.48 \text{ dam}}{1000}$
$\approx 7.16 \text{ dam}$

7. $1350 \text{ m} \cdot \dfrac{100 \text{ cm}}{1 \text{ m}} \cdot \dfrac{1 \text{ ft}}{30.48 \text{ cm}} = \dfrac{1350 \cdot 100 \text{ ft}}{30.48}$
$\approx 4429.13 \text{ ft}$

9. $0.6 \text{ in.} \cdot \dfrac{2.54 \text{ cm}}{1 \text{ in.}} \cdot \dfrac{10 \text{ mm}}{1 \text{ cm}} = \dfrac{0.6 \cdot 2.54 \cdot 10 \text{ mm}}{1}$
$= 15.24 \text{ mm}$

11. $0.06 \text{ hm} \cdot \dfrac{10,000 \text{ cm}}{1 \text{ hm}} \cdot \dfrac{1 \text{ ft}}{30.48 \text{ cm}}$
$= \dfrac{0.06 \cdot 10,000 \text{ ft}}{30.48}$
$\approx 19.69 \text{ ft}$

13. $1345 \text{ ft} \cdot \dfrac{30.48 \text{ cm}}{1 \text{ ft}} \cdot \dfrac{1 \text{ dm}}{10 \text{ cm}} = \dfrac{1345 \cdot 30.48 \text{ dm}}{10}$
$= 4099.56 \text{ dm}$

15. $2.35 \text{ km} \cdot \dfrac{1 \text{ mi}}{1.61 \text{ km}} = \dfrac{2.35 \text{ mi}}{1.61} \approx 1.46 \text{ mi}$

17. $42 \text{ dm} \cdot \dfrac{10 \text{ cm}}{1 \text{ dm}} \cdot \dfrac{1 \text{ ft}}{30.48 \text{ cm}} \cdot \dfrac{1 \text{ yd}}{3 \text{ ft}} = \dfrac{42 \cdot 10 \text{ yd}}{30.48 \cdot 3}$
$\approx 4.59 \text{ yd}$

19. $333 \text{ in.} \cdot \dfrac{2.54 \text{ cm}}{1 \text{ in.}} \cdot \dfrac{1 \text{ m}}{100 \text{ cm}} = \dfrac{333 \cdot 2.54 \text{ m}}{100}$
$\approx 8.46 \text{ m}$

Exercise Set A-3

1. $18 \text{ in.}^2 = \dfrac{18 \text{ in.}^2}{1} \cdot \dfrac{6.5 \text{ cm}^2}{1 \text{ in.}^2} = 117 \text{ cm}^2$

3. $40 \text{ m}^2 = \dfrac{40 \text{ m}^2}{1} \cdot \dfrac{1 \text{ yd}^2}{0.8 \text{ m}^2} = 50 \text{ yd}^2$

5. $32 \text{ acres} = \dfrac{32 \text{ acres}}{1} \cdot \dfrac{2.6 \text{ km}^2}{640 \text{ acres}} = 0.13 \text{ km}^2$

7. $18 \text{ ft}^2 = \dfrac{18 \text{ ft}^2}{1} \cdot \dfrac{0.09 \text{ m}^2}{1 \text{ ft}^2} \cdot \dfrac{100 \text{ dm}^2}{1 \text{ m}^2} = 162 \text{ dm}^2$

9. $3 \text{ yd}^2 = \dfrac{3 \text{ yd}^2}{1} \cdot \dfrac{0.8 \text{ m}^2}{1 \text{ yd}^2} \cdot \dfrac{100 \text{ dm}^2}{1 \text{ m}^2} = 240 \text{ dm}^2$

11. $103 \text{ km}^2 = \dfrac{103 \text{ km}^2}{1} \cdot \dfrac{640 \text{ acres}}{2.6 \text{ km}^2}$
$\approx 25,353.85 \text{ acres}$

13. $42 \text{ dm}^2 = \dfrac{42 \text{ dm}^2}{1} \cdot \dfrac{1 \text{ m}^2}{100 \text{ dm}^2} \cdot \dfrac{1 \text{ ft}^2}{0.09 \text{ m}^2} \cdot \dfrac{144 \text{ in.}^2}{1 \text{ ft}^2}$
$= 672 \text{ in.}^2$

15. $1875 \text{ in.}^2 = \dfrac{1875 \text{ in.}^2}{1} \cdot \dfrac{6.5 \text{ cm}^2}{1 \text{ in.}^2} \cdot \dfrac{1 \text{ dm}^2}{100 \text{ cm}^2}$
$\approx 121.88 \text{ dm}^2$

17. $5326 \text{ mm}^2 = \dfrac{5326 \text{ mm}^2}{1} \cdot \dfrac{1 \text{ cm}^2}{100 \text{ mm}^2} \cdot \dfrac{1 \text{ in.}^2}{6.5 \text{ cm}^2}$
$\approx 8.19 \text{ in.}^2$

19. $777 \text{ dm}^2 = \dfrac{777 \text{ dm}^2}{1} \cdot \dfrac{1 \text{ m}^2}{100 \text{ dm}^2} \cdot \dfrac{1 \text{ ft}^2}{0.09 \text{ m}^2}$
$\approx 86.33 \text{ ft}^2$

Exercise Set A-4

1. $120 \text{ g} = \dfrac{120 \text{ g}}{1} \cdot \dfrac{0.04 \text{ oz}}{1 \text{ g}} = 4.8 \text{ oz}$

3. $4823 \text{ cg} = \dfrac{4823 \text{ cg}}{1} \cdot \dfrac{1 \text{ kg}}{100,000 \text{ cg}} \cdot \dfrac{2.2 \text{ lb}}{1 \text{ kg}}$
$\approx 0.106 \text{ lb}$

5. $3 \text{T} = \dfrac{3 \text{ T}}{1} \cdot \dfrac{2000 \text{ lb}}{1 \text{ T}} \cdot \dfrac{1 \text{ kg}}{2.2 \text{ lb}} \cdot \dfrac{10 \text{ hg}}{1 \text{ kg}}$
$\approx 27,272.73 \text{ hg}$

7. $357,201 \text{ lb} = \dfrac{357,201 \text{ lb}}{1} \cdot \dfrac{1 \text{ kg}}{2.2 \text{ lb}} \cdot \dfrac{1 \text{ t}}{1000 \text{ kg}}$
$\approx 162.364 \text{ t}$

9. $5.75 \text{ T} = \dfrac{5.75 \text{ T}}{1} \cdot \dfrac{2000 \text{ lb}}{1 \text{ T}} \cdot \dfrac{1 \text{ kg}}{2.2 \text{ lb}} \cdot \dfrac{1 \text{ t}}{1000 \text{ kg}}$
$\approx 5.227 \text{ t}$

11. $213 \text{ oz} = \dfrac{213 \text{ oz}}{1} \cdot \dfrac{28 \text{ g}}{1 \text{ oz}} \cdot \dfrac{10 \text{ dg}}{1 \text{ g}} = 59,640 \text{ dg}$

13. $815 \text{ dag} = \dfrac{815 \text{ dag}}{1} \cdot \dfrac{10 \text{ g}}{1 \text{ dag}} \cdot \dfrac{0.04 \text{ oz}}{1 \text{ g}} = 326 \text{ dag}$

15. $183 \text{ oz} = \dfrac{183 \text{ oz}}{1} \cdot \dfrac{28 \text{ g}}{1 \text{ oz}} \cdot \dfrac{10 \text{ dg}}{1 \text{ g}} = 51{,}240 \text{ dg}$

17. $27 \text{ lb} = \dfrac{27 \text{ lb}}{1} \cdot \dfrac{1 \text{ kg}}{2.2 \text{ lb}} \cdot \dfrac{10{,}000 \text{ dg}}{1 \text{ kg}}$
$\approx 122{,}727.27 \text{ dg}$

19. $41 \text{ lb} = \dfrac{41 \text{ lb}}{1} \cdot \dfrac{1 \text{ kg}}{2.2 \text{ lb}} \cdot \dfrac{1000 \text{ g}}{1 \text{ kg}} \approx 18{,}636.36 \text{ g}$

Exercise Set A-5

Note: Answers can vary, depending on rounding and on the conversion factors used.

1. $3 \text{ cubic feet} = \dfrac{3 \text{ cubic feet}}{1} \cdot \dfrac{7.48 \text{ gal}}{1 \text{ cubic foot}} \cdot \dfrac{4 \text{ qt}}{1 \text{ gal}} \cdot \dfrac{2 \text{ pt}}{1 \text{ qt}} \cdot \dfrac{16 \text{ fluid ounces}}{1 \text{ pt}} = 2872.32 \text{ fluid ounces}$

3. $400 \text{ gal} = \dfrac{400 \text{ gal}}{1} \cdot \dfrac{1 \text{ cubic foot}}{7.48 \text{ gal}} \approx 53.48 \text{ cubic feet}$

5. $12{,}561 \text{ fluid ounces} = \dfrac{12{,}561 \text{ fluid ounces}}{1} \cdot \dfrac{1 \text{ pt}}{16 \text{ fluid ounces}} \cdot \dfrac{1 \text{ qt}}{2 \text{ pt}} \cdot \dfrac{1 \text{ gal}}{4 \text{ qt}} \approx 98.13 \text{ gal}$

7. $22{,}000 \text{ cubic feet}$
$= \dfrac{22{,}000 \text{ cubic feet}}{1} \cdot \dfrac{7.48 \text{ gal}}{1 \text{ cubic foot}}$
$= 164{,}560 \text{ gal}$

9. $8 \text{ cubic yards} = \dfrac{8 \text{ cubic yards}}{1} \cdot \dfrac{202 \text{ gal}}{1 \text{ cubic yard}}$
$= 1616 \text{ gal}$

11. $4.5 \text{ L} = \dfrac{4.5 \text{ L}}{1} \cdot \dfrac{1000 \text{ cm}^3}{1 \text{ L}} = 4500 \text{ cm}^3$

13. $28.5 \text{ cm}^3 = \dfrac{28.5 \text{ cm}^3}{1} \cdot \dfrac{1 \text{ L}}{1000 \text{ cm}^3} \cdot \dfrac{1000 \text{ mL}}{1 \text{ L}}$
$= 28.5 \text{ mL}$

15. $433 \text{ mL} = \dfrac{433 \text{ mL}}{1} \cdot \dfrac{1 \text{ L}}{1000 \text{ mL}} \cdot \dfrac{1000 \text{ cm}^3}{1 \text{ L}}$
$= 433 \text{ cm}^3$

17. $32 \text{ L} = \dfrac{32 \text{ L}}{1} \cdot \dfrac{1000 \text{ cm}^3}{1 \text{ L}} = 32{,}000 \text{ cm}^3$

19. 32 cL
$= \dfrac{32 \text{ cL}}{1} \cdot \dfrac{1 \text{ L}}{100 \text{ cL}} \cdot \dfrac{1000 \text{ cm}^3}{1 \text{ L}} \cdot \dfrac{1 \text{ m}^3}{1{,}000{,}000 \text{ cm}^3}$
$= 0.00032 \text{ m}^3$

21. $500 \text{ cubic feet} = \dfrac{500 \text{ cubic feet}}{1} \cdot \dfrac{62.5 \text{ lb}}{1 \text{ cubic foot}}$
$= 31,250 \text{ lb}$

23. $200 \text{ lb} = \dfrac{200 \text{ lb}}{1} \cdot \dfrac{1 \text{ cubic foot}}{62.5 \text{ lb}} = 3.2 \text{ cubic feet}$

Exercise Set A-6

1. $F = \dfrac{9}{5}C + 32 = \dfrac{9}{5}(14) + 32 = 57.2°\text{F}$

3. $F = \dfrac{9}{5}C + 32 = \dfrac{9}{5}(55) + 32 = 131°\text{F}$

5. $F = \dfrac{9}{5}C + 32 = \dfrac{9}{5}(150) + 32 = 302°\text{F}$

7. $F = \dfrac{9}{5}C + 32 = \dfrac{9}{5}(-18) + 32 = -0.4°\text{F}$

9. $F = \dfrac{9}{5}C + 32 = \dfrac{9}{5}(-33) + 32 = -27.4°\text{F}$

11. $C = \dfrac{5}{9}(F - 32) = \dfrac{5}{9}(5 - 32) = -15°\text{C}$

13. $C = \dfrac{5}{9}(F - 32) = \dfrac{5}{9}(32 - 32) = 0°\text{C}$

15. $C = \dfrac{5}{9}(F - 32) = \dfrac{5}{9}(100 - 32) \approx 37.78°\text{C}$

17. $C = \dfrac{5}{9}(F - 32) = \dfrac{5}{9}(-10 - 32) \approx -23.33°\text{C}$

19. $C = \dfrac{5}{9}(F - 32) = \dfrac{5}{9}(-14 - 32) \approx -25.56°\text{C}$

Notes

Notes

Notes

Notes

Notes

Notes

Notes

Notes